Gunnar Bruns

Electronic switching in phase-change materials

Gunnar Bruns

Electronic switching in phase-change materials

switching speed, transient phenomena, and resistance drift

Südwestdeutscher Verlag für Hochschulschriften

Impressum/Imprint (nur für Deutschland/only for Germany)
Bibliografische Information der Deutschen Nationalbibliothek: Die Deutsche Nationalbibliothek verzeichnet diese Publikation in der Deutschen Nationalbibliografie; detaillierte bibliografische Daten sind im Internet über http://dnb.d-nb.de abrufbar.
Alle in diesem Buch genannten Marken und Produktnamen unterliegen warenzeichen-, marken- oder patentrechtlichem Schutz bzw. sind Warenzeichen oder eingetragene Warenzeichen der jeweiligen Inhaber. Die Wiedergabe von Marken, Produktnamen, Gebrauchsnamen, Handelsnamen, Warenbezeichnungen u.s.w. in diesem Werk berechtigt auch ohne besondere Kennzeichnung nicht zu der Annahme, dass solche Namen im Sinne der Warenzeichen- und Markenschutzgesetzgebung als frei zu betrachten wären und daher von jedermann benutzt werden dürften.

Coverbild: www.ingimage.com

Verlag: Südwestdeutscher Verlag für Hochschulschriften GmbH & Co. KG
Heinrich-Böcking-Str. 6-8, 66121 Saarbrücken, Deutschland
Telefon +49 681 37 20 271-1, Telefax +49 681 37 20 271-0
Email: info@svh-verlag.de

Approved by: Aachen, RWTH, Diss., 2012

Herstellung in Deutschland (siehe letzte Seite)
ISBN: 978-3-8381-3319-5

Imprint (only for USA, GB)
Bibliographic information published by the Deutsche Nationalbibliothek: The Deutsche Nationalbibliothek lists this publication in the Deutsche Nationalbibliografie; detailed bibliographic data are available in the Internet at http://dnb.d-nb.de.
Any brand names and product names mentioned in this book are subject to trademark, brand or patent protection and are trademarks or registered trademarks of their respective holders. The use of brand names, product names, common names, trade names, product descriptions etc. even without a particular marking in this works is in no way to be construed to mean that such names may be regarded as unrestricted in respect of trademark and brand protection legislation and could thus be used by anyone.

Cover image: www.ingimage.com

Publisher: Südwestdeutscher Verlag für Hochschulschriften GmbH & Co. KG
Heinrich-Böcking-Str. 6-8, 66121 Saarbrücken, Germany
Phone +49 681 37 20 271-1, Fax +49 681 37 20 271-0
Email: info@svh-verlag.de

Printed in the U.S.A.
Printed in the U.K. by (see last page)
ISBN: 978-3-8381-3319-5

Copyright © 2012 by the author and Südwestdeutscher Verlag für Hochschulschriften GmbH & Co. KG and licensors
All rights reserved. Saarbrücken 2012

Acknowledgements

An interdisciplinary work like this demands for perfect interplay between different affiliations and people with different aims and different ways of thinking and working. I am grateful to all those people which have enabled the fast and successfully progressing projects of my work.

I acknowledge the opportunity to contribute to the fascinating field of phase-change material research which has been made possible by my PhD-advisor Matthias Wuttig. I would like to thank Rainer Waser, the second advisor of this work, for the fruitful projects with his department.

The custom made setups which have been build in this work have been created together with Carl Schlockermann in a close and complementary work. Philipp Merkelbach has developed and performed most of the lithographic processes. He and Peter Zalden were great office mates, and they were the first address to discuss new results and to solve problems.

The Qimonda project allowed a close connection to Jan Boris Philipp, Thomas Happ, and Michael Kund. This project was a perfect example for an industry-university joint venture and was a driving force towards the results of this thesis.

Martin Salinga and Daniel Wamwangi have been advisors in all fields of physics. Regarding phase-change switching effects and resistance drift Daniel Krebs was an indispensable discussion partner.

A close and fruitful teamwork with my three diploma students, Martin Wimmer, Karsten Fleck, and Rüdiger Schmidt, has lead to exciting experiments and results which can be found all over this work. Stephan Hermes skills and his commitment maintains our equipment, and therefore, secures a failure free work environment. The precise and fast work of our team in the mechanical and electronic workshops, especially Axel Gross, Franz Neus, and Gert Kirchhoff, were indispensable for the reliability of the custom made setups. Ralf Detemple, Sarah Schlenter, and Josefine Elbert enabled a simple and productive management of fundings, equipment acquisitions, and all other aspects of administrations. Special topics and aspects of this thesis have been discussed with Janika Boltz, Peter Jost, Andreas Kaldenbach, Michael Klein, Anja König, Stephan Kremers, Dominic Lencer, Jennifer Luckas, Hanno Volker, and Michael Woda.

I'd like to thank my family and all my friends for their support.

I

Abstract

Phase-change materials (PCM) possess a unique property contrast between their crystalline and amorphous phases. Differences in resistivity of some orders of magnitude can be observed between both phases. Also, the reflectivity shows a remarkable contrast, and it allows the application of thin phase-change layers in optical media, such as CD, DVD and Bluray disk, to enable rewritable storage of information. Although both phases are stable for decades at room temperature, it is possible to switch between the phases in nanoseconds at elevated temperatures. This striking combination of stability and rapid transition, together with the pronounced resistivity contrast, make PCMs one of the most promising candidates for future, non-volatile, electronic memory. In this work, three physical aspects concerning such a memory have been investigated using custom made setups to cope with the challenges of subnanosecond timescales and resistances of hundreds of Gigaohms.

Speed limitations of write and rewrite operations are a crucial topic, if PCMs shall be able to compete with the established concepts in electronic devices. Two classes of storage devices are used in modern computer systems, so far. On the one hand, there is the fast, but volatile memory close to the processor unit, like the dynamic (DRAM) and the static (SRAM) random access memory. On the other hand, there are slower, but non-volatile storages, like hard disk drive, flash, and optical media. In terms of speed, there is a gap of several orders of magnitude between memory and storage concepts. Phase-change memory could close this gap and establish a new storage class memory, and maybe, even allow to build a non-volatile memory device, which could replace the volatile DRAM. Investigations regarding the memory switching speed in phase-change memory will be presented in this work, and reveal that the phase transitions can be accomplished within a few nanoseconds. This demonstrates the potential of PCMs to compete with DRAM in terms of speed.

The second topic in this work are transient phenomena, like threshold switching, which occur when PCMs are treated with electrical pulses. Threshold switching describes a sudden decrease of the material's resistivity. This effect can be observed in amorphous PCMs, if the applied electrical field exceeds a threshold value. In this work, results will be presented which describe both the resistance drop during the threshold switching and the life time of this high conductive state. While present publications use a characteristic field strength for each PCM to describe the occurrence of the threshold switch, the results of this work suggest the definition of a field dependent delay time, which predicts the sudden change of conductivity.

Besides the extraordinary behavior of disordered semiconductors at high electric fields, there is a further effect at low fields, which has a tremendous influence on phase-change memory applications: the resistance drift. This effect describes the time dependent increase of the resistivity of amorphous PCMs. In this work, experimental data will be presented which demonstrate the similarity of this effect in both unstructured films and memory devices. These data have been used to modify an existing model which explains the drift's origin. The reported dependency of the drift behavior on the activation energy for conduction can be confirmed and a new aspect of this dependency will be presented.

Kurzfassung

Übersetzung des englischen Originaltitels: Elektronisches Schalten in Phasenwechselmaterialien.

Phasenwechselmaterialien (PCM) besitzen einen einzigartigen Kontrast physikalischer Eigenschaften zwischen ihrer kristallinen und amorphen Phase. So unterscheidet sich der spezifische Widerstand beider Phasen in manchen Materialien um mehrere Größenordnungen. Ebenso gibt es einen beachtlichen Kontrast im Reflexionsvermögen, der bereits industrielle Anwendung gefunden hat: in optischen Medien wie CD, DVD und Bluray Disks ermöglichen dünne Schichten aus PCM eine mehrfach wiederbeschreibbare Datenspeicherung. Und obwohl beide Phasen an Raumtemperatur für Jahrzehnte stabil sind, ist es bei höheren Temperaturen möglich, innerhalb von Nanosekunden zwischen den Phasen zu wechseln. Diese erstaunliche Kombination von Stabilität und schnellen Übergängen, gemeinsam mit dem ausgeprägten Eigenschaftskontrast, macht PCMs zu einem vielversprechenden Kandidaten für zukünftige, nicht flüchtige, elektronische Speicher. In dieser Arbeit wurden drei physikalische Aspekte solcher Speicher untersucht, wofür speziell angefertigte Messplätze geschaffen wurden, um die Herausforderungen zu meistern, die bei Messungen auf der Picosekunden-Skala und mit Widerständen von mehreren hundert Gigaohm auftreten.

Die Limitierung der Geschwindigkeit von Schreib- und Lösch-Operationen ist ein wesentliches Thema bei der Realisierung auf PCMs basierender elektronischer Speicher, die mit den bereits etablierten Konzepten konkurrieren sollen. Zur Zeit werden zwei Klassen zur Speicherung elektronischer Daten in Computern genutzt. Zum einen gibt es die schnellen, aber flüchtigen Arbeitsspeicher (memory), wie DRAM (dynamic random access memory) und SRAM (static random access memory). Zum anderen sind da die langsameren, aber nicht flüchtigen Speicher (storage), wie Festplatte, Flash und optische Medien. Unter dem Blickwinkel der Geschwindigkeit klafft eine Lücke von mehrerer Größenordnungen zwischen den Konzepten Arbeitsspeicher und Langzeitspeicher. Phasenwechsel-Speicher könnten diese Lücke schließen und eine neue Speicherklasse (storage class memory) schaffen und eventuell sogar einen nicht flüchtigen Arbeitsspeicher ermöglichen, der den flüchtigen DRAM ersetzt. In dieser Arbeit werden Untersuchungen der Schaltgeschwindigkeiten in Phasenwechsel-Speichern präsentiert, die belegen, dass der Phasenübergang innerhalb weniger Nanosekunden stattfinden kann. Dies demonstriert das Potential der PCMs unter dem Gesichtspunkt der Schaltgeschwindigkeit mit dem DRAM mithalten zu können.

Das zweite Thema dieser Arbeit sind die Übergangsphänomene, die auftreten, wenn PCMs hohen elektrischen Feldern ausgesetzt werden. Das „Threshold switching" beschreibt die plötzliche Abnahme des spezifischen Widerstandes. Dieser Effekt kann in amorphen PCMs beobachtet werden, sobald die angelegten Feldstärken einen Schwellenwert überschreiten. In dieser Arbeit werden Ergebnisse präsentiert, die sowohl den Einbruch des Widerstandes mit dem „Threshold switching" beschreiben, als auch die Lebenszeit des damit verbundenen angeregten Zustandes hoher Leitfähigkeit. Während aktuelle Publikationen eine charakteristische Feldstärke für jedes PCM nutzen, um das Auftreten des „Threshold switching" zu beschreiben, kann aus den Ergebnissen dieser Arbeit gefolgert werden, dass es eine feldabhängige Verzögerungszeit gibt, die die plötzliche Änderung der Leitfähigkeit vorhersagt.

Neben diesem außergewöhnlichen Verhalten un-

III

geordneter Halbleiter unter dem Einfluss hoher elektrischer Felder gibt es einen weiteren Effekt, der auch ohne äußere Felder auftritt und einen enormen Einfluss auf Phasenwechsel-Speicher hat: die Widerstandsdrift. Dieser Effekt beschreibt das zeitliche Ansteigen des spezifischen Widerstandes amorpher PCMs. In dieser Arbeit werden Daten präsentiert, die die Gleichartigkeit dieses Effektes in unstrukturierten Filmen und in Speicherzellen belegen. Diese Daten wurden genutzt, um ein existierendes Modell zu modifizieren, das den Ursprung der Drift erklärt. Die bekannte Abhängigkeit des Drift Verhaltens von der Aktivierungsenergie für den Ladungstransport kann bestätigt werden und zusätzlich kann ein neuer Aspekt dieser Abhängigkeit gezeigt werden.

Contents

1	**Phase-Change Memory**	**1**
	1.1 Phase-Change Materials	2
	1.2 Electronic Memory	6
2	**Theoretical Models for Memory and Threshold Switching**	**13**
	2.1 Phase Transitions	13
	2.1.1 Crystallization: Nucleation	14
	2.1.2 Crystallization: Growth	16
	2.1.3 Amorphization: Glass Formation	16
	2.2 Threshold Switching	17
	2.2.1 Generation and Recombination of Carriers	18
	2.2.2 Electric Field Induced Nucleation	21
3	**Experimental Methods**	**23**
	3.1 Phase-Change Memory Cells	23
	3.1.1 Sample Processing	24
	3.1.2 Challenges with State-of-the-Art Nanodevices	28
	3.2 PET: Pulsed Electrical Tester	28
	3.2.1 Setup Description	31
	3.2.2 Software and Data Analysis	33
	3.2.3 Calibration and Testing	34
	3.2.4 Pulse Parameter Definition	35
	3.3 Sheet Resistance Measurements - van-der-Pauw Setup	36
	3.3.1 Setup Description	36
4	**Phase Transition in Memory Devices**	**41**
	4.1 Initialization Procedures in Mushroom Cells	41
	4.1.1 Cell Training	42
	4.1.2 Recreate Initial State	42
	4.2 Nanosecond Switching in GeTe Memory Cells	43
	4.2.1 Experimental Parameters	44
	4.2.2 Crystallization Window	45
	4.2.3 Growth Dominated Recrystallization	46
	4.2.4 Competing with DRAM Speed	49
	4.3 Switching Experiments with GST	52
	4.4 Competing Material: AIST	55
5	**Transient Phenomena**	**59**

	5.1	Threshold Switching Delay Time	59
		5.1.1 Rectangular Pulse Experiment	60
		5.1.2 Leading Edge Variation	63
		5.1.3 Field Dependent Threshold Time	64
	5.2	Switching Timescales	67
	5.3	Excitation Livetime	68
		5.3.1 Experiment Design	68
		5.3.2 Threshold Voltage Recovery Time	71

6 Resistance Drift in Amorphous as Deposited Phase-Change Films   **75**
    6.1 Theoretical Background 76
    6.2 Comparison with Experiments in Memory Devices 77
    6.3 Dependence of Drift Energy on Temperature 79
    6.4 Dependence of Drift Energy on Activation Energy for Conduction 82

7 Conclusion   **85**

Bibliography   **89**

List of Figures

1.1	Sketch of Memory and Storage Concepts	3
1.2	Sketch of Atomic Configuration in Different Phases	4
1.3	Idealized Temperature Dependent Resistance Measurement	5
1.4	Principle of PCM Devices	7
1.5	Principle of Phase-Change Storage Devices	9
1.6	Principle of Multilevel Storage	10
2.1	Crystallization: Nucleation and Growth	16
2.2	Threshold Switch in Memory Cells	18
2.3	Energetic Band Diagram for Generation Model	20
2.4	Electron Potential for Hopping Transport	21
3.1	PET Sample Cross Section	24
3.2	Qimonda Samples	25
3.3	Lithography and Lift-Off Process	27
3.4	AFM Picture of Bottom Electrode	28
3.5	Picture of the PET Contact Board	29
3.6	PET Schematic	32
3.7	PET: Shortest Possible Pulses	35
3.8	Picture of the van-der-Pauw Setup	37
3.9	Sample Geometrie for van-der-Pauw Setup	38
3.10	Schematic of the van-der-Pauw Setup	39
4.1	Pulse Pattern for Cell Training	43
4.2	Pulse Pattern to Test Crystallization Speed	44
4.3	GeTe PTE - Crystallization Window	45
4.4	Crystallization in Phase-Change Devices	47
4.5	GeTe PTE - Growth Dominated	48
4.6	SET Times in Memory Cells	50
4.7	Nanosecond Switching in GeTe	51
4.8	$Ge_2Sb_2Te_5$ PTE	53
4.9	AgIn-doped Sb_2Te PTE	56
5.1	Threshold Switching Delay - Voltage Pulses	60
5.2	Threshold Switching Delay Time - Dependency on Cell Resistance	61
5.3	I-V-Curves: Threshold Switch	63
5.4	Calculated and Measured Threshold Voltage	66
5.5	Threshold Switch Livetime: Excitation Pulse	69
5.6	Threshold Switch Livetime: Probe Pulse	69

5.7	Threshold Switch Livetime: I-V-Curves	71
5.8	Threshold Voltage Recovery Time	73
6.1	Resistance Drift of GeTe	79
6.2	Drift Energy of GeTe	81
6.3	Drift Energy and Activation Energy of Phase-Change Materials	82

List of Tables

4.1	Crystallization and Melting Temperatures	54
4.2	Electrical Properties of Phase-Change Materials	55
6.1	Drift Coefficients of GeTe	78
6.2	Drift and Activation Energy of Phase-Change Materials	84

1 Phase-Change Memory

The storage and preservation of knowledge is the foundation of our civilization. Only non-volatile data storage enables to accumulate knowledge and to provide it to future generations. During the centuries, the methods to store information have been improved in terms of reliability and usability. From symbols, carved in stone, and letters, written on paper, it was a long way to electronic data storage.

The technologies to handle information within a computer have split in two branches: fast, but volatile memories, and slower, but non-volatile storage concepts. Memory devices allow very fast random-access rewrite and read processes, therefore, they are suitable as cache and main memory close to the central processor unit (CPU). Operational speeds of some nanoseconds are comparable with the processor frequency and have been made possible by using CMOS[1] transistors, in the SRAM[2] concept, and small capacitors, in the DRAM[3] concept. A major disadvantage of both concepts is the volatility of the stored information. The transistor concept in SRAM necessitates a constant application of a supply voltage to maintain the transistor state. In the DRAM concept, the charge of a capacitor represents the binary information, and the leak currents within this capacitor demand for repetitive refreshing.

In contrast to this purely electronic memory concepts, there are many types of devices used for long term storage of information. In hard disc drives (HDD) and magnetic tapes, the electronic information is converted to different orientations of magnetic domains in ferro-magnetic materials. Although the lifetime of the stored data can be some decades, the access times for rewrite and read processes are in the range of milliseconds, or higher, and orders of magnitude slower compared to the timescales in electronic memories. Optical media, like CD, DVD, and Blu-ray discs, utilize the reflectivity contrast in polymers or phase-change alloys [Yamada:1987] to store information, and their access times are comparable with the timescales of magnetic devices.

A bridge between volatile electronic memories and non-volatile storage devices was built by introducing the flash memory [Pashley:1989]. Comparable to the DRAM concept, the information is stored using a small capacitor, but in contrast to DRAM, the flash capacitor has a

---
[1]Complementary metal-oxide-semiconductor.
[2]Static random-access memory. Fastest memory concept, used as cache at the CPU.
[3]Dynamic random-access memory. Highest possible storage density, used in main memory.

Chapter 1

floating gate, which is isolated by a thin dielectric layer. High electric fields force a tunneling of electrons to the floating gate, where they remain, until removed by even higher electric fields. The access times of flash devices are in between of those of storage and memory concepts. Read out operations can be performed within some microseconds, but writing, and especially deleting, of data is comparable with HDDs.

Phase-change memory possesses the ability to combine non-volatility and fast access times to found the intermediate type of storage class memory [Raoux:2008b]. The fast atomic rearrangements in so called phase-change materials, combined with the large resistivity contrast between amorphous and crystalline phases in this material class, can be used to store information for several years [Pirovano:2004], with access times of a few nanoseconds [Bruns:2009].

The research field concerning the electronic properties of phase-change materials has grown constantly over the last decade. Many review papers have been published to summarize and conclude the increasing amount of experimental observations and theoretical models [Bez:2005, Welnic:2008, Lai:2008, Raoux:2008b, Waser:2010, Raoux:2010, Burr:2010, Salinga:2011]. A comprehensive work is [Burr:2010], in which the author also presents aspects of the history of the development of phase-change memory and provides descriptions of the key properties of competing technologies (Fig. 1.1).

Discussions of the most important theoretical models, regarding switching operations in phase-change memory, will be presented in chapter 2 followed by experimental results in chapters 4, 5, and 6, which are summarized in chapter 7. But at first, the following sections will introduce the remarkable properties of phase-change materials, and explain the working principle of phase-change memory.

1.1 Phase-Change Materials

Phase-change materials obtained their name from their characteristic behavior upon phase transformations. With the change of the long range order, between an amorphous and a crystalline phase, this material class possesses an unusual high contrast of physical properties, like dielectric constants and resistivity. The last one will be the most important material property in this work, and most of the presented models and experiments will focus on changes in the materials' resistivities.

In addition to the pronounced property contrast between the phases, the kinetics of the phase transition is remarkable. Despite the fast observed crystallization within nanoseconds [Bruns:2009], at elevated temperatures, the amorphous phase is stable at room temperature for decades.

Materials, which possess this unique combination of property contrast and high dynamic

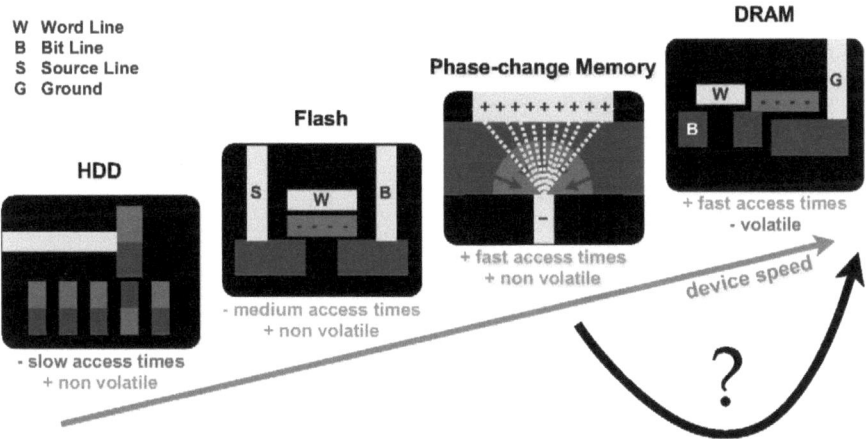

Figure 1.1: Sketch of memory and storage concepts arranged in order maximum read and write speeds. Hard disk drives (HDD) utilize switching of magnetic domains to store information, while flash and DRAM capture electrons in capacitors. In phase-change memory (PCRAM) the high resistivity contrast between crystalline and amorphous phase is used to store information. In terms of speed, the non-volatile PCRAM has the potential to compete with the volatile DRAM.

range of crystallization speed, have been classified as phase-change materials. Recently, a number of publications have revealed fundamental structural characteristics, which phase-change materials have in common. The interplay of local structure and physical properties [Welnic:2006] was investigated and lead to an explanation of the origin of the optical contrast in phase-change materials [Welnic:2007]. Consequently, the resonant bonding in crystalline phase-change materials [Shportko:2008] was identified to be responsible for the high dielectricity in the crystalline state, in contrast to the amorphous state. To enable resonant bonding, two aspects of bonding properties have to be considered. Both, the ionicity and the hybridization of the candidate alloys, have to be in a certain range [Lencer:2008]. A map, using those two properties as coordinates, allows to predict which alloys have the potential to show resonant bonding in their crystalline phase, and therefore, can be good phase-change materials.

In a simplified picture (Fig. 1.2), there are three characteristic states necessary to explain the phase-change processes: crystalline, amorphous and molten. The kinetics of the required phase transformations will be discussed in chapter 2.1 in detail. In this section, only a brief overview will be presented to define the nomenclature and allow an explanation of the working principle of phase-change memory devices in the following section 1.2.

Chapter 1

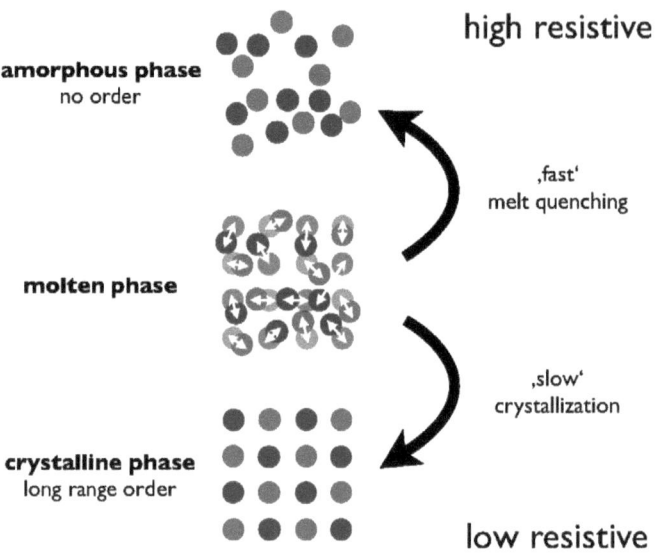

Figure 1.2: Sketch of atomic configurations in different phases. Slow cooling of the molten material allows ordering of the atoms in the energetic favorable crystalline phase. Fast quenching of the molten state will "freeze" the liquid in the unordered state.

Below the melting temperature T_{melt}, the crystalline phase is energetically favorable. In the crystalline phase, the atoms are arranged in a periodic pattern, which leads to a metallic conduction mechanism, if the disorder of the crystal becomes small enough [Siegrist:2011]. However, amorphous phase-change materials will not crystallize at room temperature, because of their very high viscosity. The high amount of disorder in amorphous phase-change materials prohibits metallic conduction, but it creates a high amount of localized trap states.

A hopping transport describes the conduction mechanism in the amorphous phase [Ielmini:2007], and therefore, a thermally activated conduction can be observed. Hence, the high resistance R of an amorphous sample will decrease on annealing, following an Arrhenius correlation (Fig. 1.3):

$$R(T) = R_x \cdot e^{\frac{E_A}{k_B T}}, \tag{1.1}$$

where E_A is the activation energy for conduction and R_x is a phenomenological prefactor with the dimensions of a resistance. Increasing the temperature will lower the activation barrier

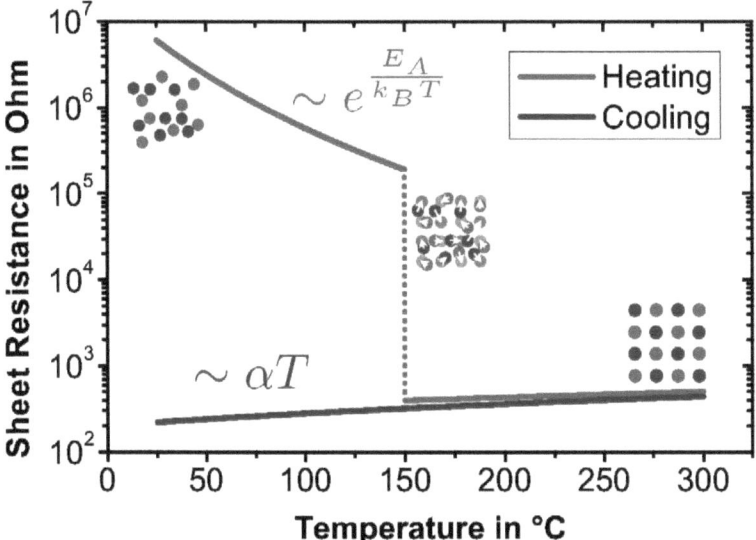

Figure 1.3: Idealized temperature dependent resistance measurement. Thermal activated carriers increase the conductivity, and therefore, decrease the resistance of the amorphous phase on annealing. Reaching T_{crys}, the resistance plummets, due to the phase transition, some orders of magnitude, and the slope of the measurement changes, and reveals the metallic conduction mechanism characteristic for the crystalline state of many phase-change materials.

for the crystallization process, and leads finally, when the crystallization temperature T_{crys} is reached, to a fast and complete crystallization of the phase-change material. Conjunct with the crystalline order, the conduction mechanism changes, and the resistivity decreases. Some phase-change materials, like germanium telluride (GeTe) and doped antimony telluride (Sb_2Te), crystallize in a sudden and distinct way causing the resistance of an amorphous sample to plummet orders of magnitude within some seconds, as soon as T_{crys} is reached (Fig. 1.3). Other materials, like $Ge_2Sb_2Te_5$, show a more smooth and more complicated transition (see [Friedrich:2000]), which can be explained with a disorder induced metal-insulator transition (MIT) [Siegrist:2011].

Further heating of the sample will increase the resistance, due to the metallic conduction mechanism:

$$R(T) = R_0 \cdot [1 + \alpha(T - T_0)], \qquad (1.2)$$

where α is the temperature coefficient of resistivity and R_0 is the resistance at the temperature T_0. Cooling the sample back below T_{crys} will not change the crystalline state. Therefore, the resistance will decrease according to formula 1.2, and at room temperature, the resistance of the sample will be orders of magnitude smaller than its initial resistance in the amorphous phase (Fig. 1.3).

This contrast in resistivity can be utilized in electronic memory devices. The switching, from amorphous to crystalline state, can be evoked by heating of the phase-change material by an electric current. For the amorphization process, the phase-change material has to be molten and quenched rapidly to prohibit a crystallization during cooling. The theoretical concepts, necessary to understand this behavior, will be discussed in chapter 2, but at first, the working principle of a phase-change memory cell will be explained.

1.2 Electronic Memory

Despite the simplicity and elegance of the basic idea, how to build a phase-change memory cell, there are many challenges on the way to realize one. Already in 1970, Neale et al. proposed an electronic memory, based on phase-change materials, which uses electronic pulses to heat and switch the material between its crystalline and amorphous phase [Neale:1970]. To read the stored information, voltage pulses can be used to sense the resistance of the material. Therefore, a fully electronic memory element would be realized, which can be used in a computer or in comparable applications.

In the beginning, two physical aspects have to be taken into account. First of all, if the phase-change material shall be switched using electronic pulses, the design of the memory cell should allow high current densities to enable temperatures, high enough, to melt the material with the evoked Joule heat. Secondly, the cell design must allow very fast cooling of the phase-change material to enable a melt-quench process, otherwise, the material will always crystallize during the cooling, and the amorphous phase could not be formed. Both aspects demand for a cell geometry of very small dimensions.

Regarding the required high current densities, some technological constraints have to be taken into account. With focus on future applications of phase-change memory in mobile devices, the voltage heights of the switching pulses should be below 3.7 V, which is the supply voltage of the commonly used lithium polymer battery in mobile applications. To achieve current densities which allow local melting of phase-change material using these low voltages the electrode size in a phase-change memory cell has to be tailored to the typical resistivities and thermal properties of electrode and phase-change materials. Therefore, in case of standard materials like $Ge_2Sb_2Te_5$, the necessary diameter of the electrodes has to be less than 100 nm.

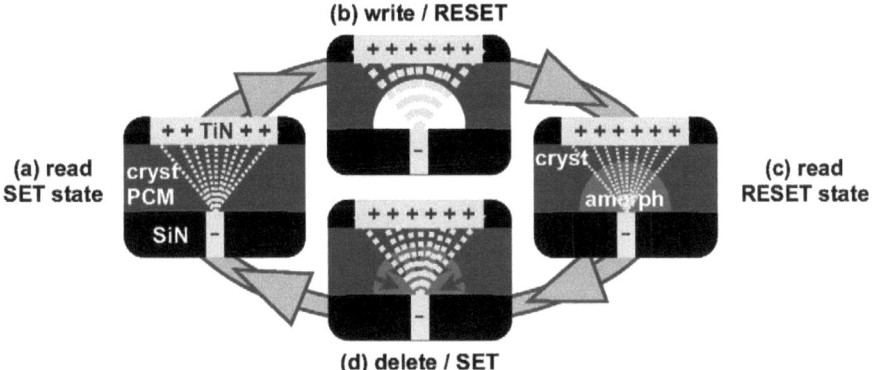

Figure 1.4: Working principle of a phase-change memory cell, shown in a cross section. The phase-change material (PCM) is sandwiched between a large area top electrode (titanium nitride, TiN) and a small bottom electrode (TiN) in a dielectric (silicon nitride, SiN). SET and RESET state are written with a moderate (b), or a high voltage pulse (d), respectively. Both can be easily sensed with a low pulse (a) & (c).

In addition, cooling rates of 100 K/ns are required for the melt-quenching to the amorphous phase. Hence, the volume of the phase-change material has to be small in comparison to the contact surface to the surrounding dielectric material.

Simplified cross sections of phase-change memory cells in mushroom design are shown in Fig. 1.4. Each cell consists of four main components: two electrodes, a phase-change layer, and a surrounding dielectric. The phase-change material is sandwiched between two metallic electrodes. While the top electrode has a large contact area to the phase-change material, the bottom electrode has a very small diameter of less than 100 nm.

Application of a voltage at the electrodes leads to a current through the phase-change material. Due to the different contact areas, of top and bottom electrodes, with the phase-change material (PCM), the current density will be much higher in the bottom electrode and at the interface with the PCM. Therefore, the bottom electrode is the source of the Joule heating. This has motivated the name *heating electrode*, or simply *heater*, for the bottom electrode.

The ground level state of a cell is called the SET state, in which the PCM is in its crystalline phase, this state is the logical "0" of the memory. To store a logical "1" in one cell, a part of the PCM volume has to be amorphized to create the RESET state. Recrystallizing this amorphous volume will delete the "1" and restore the SET state. Therefore, the deleting process is called SET operation, while the amorphization process is the RESET operation. These nomenclature was adapted from write and delete terms in optical memory disks.

Chapter 1

To RESET a cell, a high voltage pulse has to be applied leading to a high current between the electrodes in the order of 1 mA. This current induces enough Joule heating at the heater to melt the PCM partially. Due to the radial symmetry of the cell, the molten volume is shaped as a hemisphere above the heater. Depending on the height of the applied current, the radius of the molten volume will change due to the temperature gradient in the cell (compare [Redaelli:2005b, Russo:2009]). To SET the cell, this amorphous volume has to be recrystallized by application of a moderate voltage pulse to heat the volume above T_{crys}, but below T_{melt}, to enable the growth of the crystalline order from the surrounding material into the amorphous volume.

By using voltage pulses of much smaller height, the two cell states, SET and RESET, can be distinguished electronically. The high contrast in resistivity between the amorphous and the crystalline phase allows a very simple read out process of the cell state. Even a small amorphous volume above the heating electrode will increase the resistance of the cell by orders of magnitude. Therefore, the measurement of the current, induced by a low voltage pulse, allows to determine, wether the cell is in its SET or in its RESET state.

To summarize the properties of PCMs and storage concepts, a typical write and read cycle in a phase-change memory is visualized in Fig. 1.5, for both optical data storages and electronic memories.
Read: The initial SET state of the device can be sensed by using a low voltage pulse or a weak laser pulse. The crystalline material has a small resistivity, and therefore, the induced current is in the order of 10 µA, indicating the SET state. In optical media, the crystalline order leads to a high reflectivity, characteristic for the SET state.
RESET: Application of a high voltage pulse or an intense laser pulse melts the PCM, the subsequent fast heat dissipation of the surrounding material leads to a quenching of the molten phase, and creates an amorphous volume. This operation switches the cell to the RESET state.
Read: The resistivity of the disordered, amorphous material is orders of magnitude higher than the resistivity of the crystalline phase. Therefore, for a cell in the RESET state, the evoked current of a low voltage pulse is very small, i.e. in the order of some 100 nA. In optical media, the low reflectance of the disordered phase is detected by a weak laser pulse.
SET: To delete the written information, the cell is switched back to the SET state by application of a moderate current or laser pulse which leads to a recrystallization of the amorphous volume.

This switching from SET to RESET state and vice versa is known as memory switching. In chapter 4 the memory switching of representative phase-change materials will be characterized with focus on the necessary time for the SET process.

Phase-Change Memory

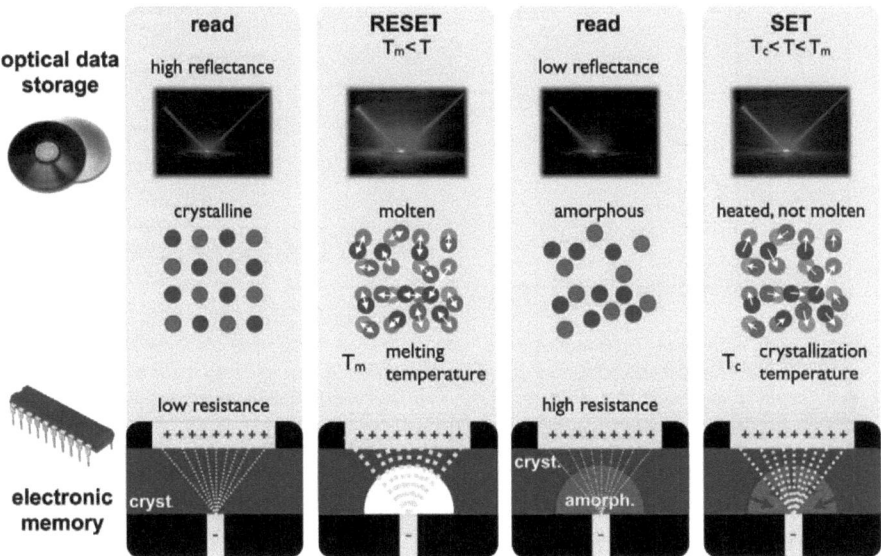

Figure 1.5: Principle of a read and write cycle in optical and electrical phase-change storage devices. Read operations: weak laser or current pulses can sense the difference in reflectance or resistance between crystalline and amorphous state. Write operations: pulses of high intensity will melt the material and quenching provokes amorphization, moderate pulses heat the material above T_{crys} and lead to recrystallization.

Besides the mushroom cell design, which has been presented as a fully functional prototype by Lai and Lowrey [Lai:2001], there are further cell concepts. A very different approach is called line cell design [Lankhorst:2005, Chen:2006]. Instead of a heating electrode the phase-change material itself provides the smallest cross section for the current flow. Hence, the center of a thin and long line of phase-change material is the source of the Joule heating, and therefore, the electrode material becomes less important, and the electrodes play a minor role in the switching process. The production steps towards line cell phase-change memory cells and investigations of their characteristics can be found in [Krebs:2009, Krebs:2009b, Krebs:2010].

Mushroom and line cells are the two basic concepts of phase-change memory cells. Further cell design can be seen as geometrical variations, or hybrids, of these concepts. E.g. the pore cell [Pirovano:2003] was first created as a modification of the mushroom cell, where the material of the heating electrode was substituted by phase-change material. And, although the resulting cell is nearly identical to the original mushroom cell, the working principle of this cell is more comparable to a line cell.

Chapter 1

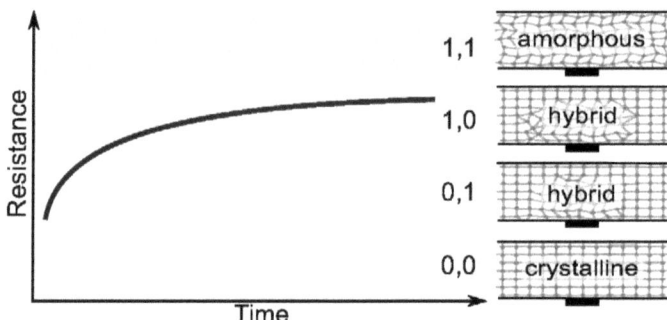

Figure 1.6: Principle of multilevel storage in phase-change storage devices. Four logic states are stored in one physical cell using four resistance windows. A possible reason for data corruption and failure is the resistance drift which increase the programmed resistance with time and change the stored information. (graphic from [Schmidt:2010])

The resistivity of most crystalline PCMs is so small that the resistance of a cell in its SET state (around $10\,\text{k}\Omega$) is dominated by the geometry and material properties of the electrodes. On the contrary, the resistivity of the amorphous phase is so high that the radius of the amorphized volume of a RESET state dominates the cell resistance (around $1\,\text{M}\Omega$). Therefore, different RESET states, with different radii of amorphous volume, can be distinguished easily by measuring the cell resistance. This allows to create a multi level storage, in which more than one logical bit is stored in a single physical memory cell [Nirschl:2007, Bedeschi:2009]. In contrast to multi level storage in flash memory [Jung:1996], the large resistivity contrast in phase-change material would allow to distinguish between more than four cell states, and therefore, storage of more than two bits in a single cell.

An example for a two-bit concept is shown in Fig. 1.6. The SET state, with the complete crystalline phase-change layer, is defined as "0,0". The maximum RESET state, with a complete amorphous phase-change layer, corresponds to the logical "1,1" state. In between, there are two hybrid states, with amorphous volumes of different size, representing the logical states "0,1" and "1,0".

The realization of a multi level phase-change memory is hampered by the so called resistance drift effect. This effect will be one of the major topics of this work, and will be discussed in chapter 6. As a motivation for the industrial importance of a profound physical investigation of this effect, the influence on a phase-change memory will be sketched: To switch a multi level cell to the "0,1" state the cell resistance has to be set to a defined resistance window (yellow resistance window in Fig. 1.6). With time the resistance of this state will increase due to the resistance drift. After a certain time span, the resistance of the cell will reach the next higher resistance window, and therefore, a read out operation would now sense the state "1,0" instead of "0,1". As consequence the stored data will be misinterpreted, and the original information is lost.

Recently, Papandreou et al. have published a method to avoid drift related information corruption [Papandreou:2011]. They utilize phenomenological derived formulas which characterize the change of the resistance of an amorphized memory cell with time. Therefore, they can expand the time span in which it is possible to distinguish between different hybrid states. This allows the industrial production of a multi level phase-change memory which preserves the stored information for a couple of weeks, but the physical origins of the drift effect is still unknown, and longterm storage is still impossible. In chapter 6, the physical models, published so far, will be discussed, and it will be shown how one of the models could be improved using the results obtained from a custom made setup and a modified experimental technique.

The third topic of this work is correlated with the conduction mechanism in amorphous phase-change materials. As already discussed, the resistivity of PCMs in their amorphous phase is orders of magnitude larger than in their crystalline phase. This would prohibit the realization of a phase-change memory switched by low voltage pulses. High voltages, in the range of some hundreds of Volts, would be necessary to evoke high current densities in memory cells, because the resistance of the RESET state is in the range of several Megaohms.

Fortunately, the threshold switching effect, discovered in the late 1960's [Ovshinsky:1968], provides a simple mechanism to utilize the conduction characteristics of disordered semiconductors for electronic memories. High electric fields increase the conductivity of amorphous phase-change materials suddenly, and drop the cell resistance to a value comparable with the resistance of the SET state. Theoretical models, describing and explaining this effect, will be discussed in chapter 2.2, and experimental data, obtained from investigations of memory cells using a custom made setup (chapter 3), will be presented in chapter 5.

The physical aspects of phase-change materials in the new storage class memory will be the topic of this work. In the following chapter 2, the theoretical models concerning kinetic and electronic properties of phase-change materials will be discussed. Subsequently, the custom made experimental tools, which fulfill the requirements to investigate these properties, will be described in chapter 3. The focus of the following experiments will be on the kinetics of phase transitions in memory devices upon application of voltage pulses (memory switching, chapter 4), followed by a detailed investigation of the electronic effect of threshold switching (chapter 5). Finally, in chapter 6, experimental and theoretical results regarding the effect of resistance drift in disordered semiconductors will be presented. The accomplishments and the major conclusions will be summarized in chapter 7.

2 Theoretical Models for Memory and Threshold Switching

The three main topics of this work are memory switching, threshold switching, and resistance drift. While both memory and threshold switching require the application of electrical pulses that induce high electric fields in the material, the resistance drift can be investigated using low field strengths. This is correlated with the nature of these phenomena. On one hand, the switching processes have to be induced by strong external excitations, while drift seems to be a material intrinsic property which can be investigated without further stimulation. On the other hand, the physical models describing the effects investigated are known and established in case of memory switching. Theoretical models for threshold switching and resistance drift, on the contrary, are under development, and do not predict certain effects. These models rather describe known effects from a phenomenological point of view.

Therefore, section 2.1 will describe the established models concerning crystallization and glass formation to explain the response of phase-change materials in memory cells upon stimulation, chosen in the experiments, which are presented in chapter 4.

Section 2.2 presents three competing models which describe the threshold switching supported by published experimental data. Chapter 5 will increase the database for these models, and presents new findings which have to be taken into account for future improvements of these models.

Subsequent to the chapters concerning memory and threshold switching, the topic of chapter 6 will be both the presentation of theoretical models of resistance drift and the experimental data obtained from the custom made setup for temperature dependent sheet resistance measurements (see chapter 3.3).

2.1 Phase Transitions

While the property combination of phase-change materials is highly unusual (compare chapter 1.1), the kinetics of the transition between the amorphous and the crystalline phase can be described with the same established models also employed for other material classes. In

Chapter 2

the following subsections, the broad spectrum of theory of crystallization kinetics presented in [Burke:1965, Porter:1992] has been condensed to summarize the theoretical background which is necessary to understand both the crystallization of an amorphous volume (section 2.1.1 and 2.1.2) and the glass formation in an undercooled liquid (section 2.1.3), which allows amorphization of crystalline volumes by melt-quenching.

The driving force of crystallization is the Gibbs free energy G

$$G = U + pV - TS = H - TS, \qquad (2.1)$$

with internal energy U, pressure p, volume V, temperature T, entropy S, and enthalpy H. The different states of phase-change material can be distinguished by their Gibbs free energy G. The differences in G can be correlated with differences in the local configuration of the atoms. The crystalline state which is described by the lowest G is the stable state. Besides this state, there are meta stable states as the amorphous phase or some crystalline phases. A rearrangement of the atoms of a meta stable state to minimize G until the stable state is reached requires breaking of inter-atomic bonds. This can be described by an energetic activation barrier which inhibits the spontaneous crystallization of the meta stable states. Above the melting temperature T_{melt} the liquid phase is the energetic favorable state. Below T_{melt} the crystallization can be described by the nucleation and growth model.

2.1.1 Crystallization: Nucleation

Starting from an amorphous phase the crystallization process has to start by building crystalline nuclei. The formation of a crystalline nucleus will decrease G due to the energetically favorable atomic arrangement within the nucleus, but the creation of an interface between the crystalline nucleus and the amorphous surrounding will increase G.

The free energy necessary to build a nucleus can be calculated for a spherical nucleus of the radius r. The volume of this nucleus will decrease G proportional to the difference ΔG_v of the molar free energy per volume. The surface of this nucleus will increase G proportional to the specific interface energy σ. In summary, the change of free energy due to the formation of a nucleus can be written as

$$\Delta G_{nucleus} = \Delta G_v \cdot \frac{4}{3}\pi r^3 + \sigma \cdot 4\pi r^2. \qquad (2.2)$$

This formula shows a local maximum

$$\Delta G_c = \frac{4}{3}\pi r_c^2 \sigma \qquad (2.3)$$

for a critical radius r_c

$$r_c = \frac{-2\sigma}{\Delta G_v}. \qquad (2.4)$$

A nucleus with a radius smaller than r_c is unstable, because the free energy will increase if a further atom is added to the nucleus. The nucleus becomes stable as soon as its radius reaches r_c. The free energy will decrease, if further atoms are added to the nucleus. This leads to a energetically driven growth of the nucleus and leads to crystallization of the whole amorphous volume (see next section 2.1.2).

The possibility to form a nucleus depends on the necessary energy $\Delta G_{nucleus}$ and can be described by a Boltzmann distribution. Therefore, the number N_c of nuclei larger than r_c depends on the number of possible nucleation centers N_0 multiplied with the possibility

$$N_c = N_0 \cdot e^{-\frac{\Delta G_c}{k_B T}} \qquad (2.5)$$

The formation and decay of crystalline nuclei in an amorphous volume can be described by the nucleation model of Vollmer and Weber [Vollmer:1925] and Becker and Döring [Becker:1935], respectively. They calculate the nucleation rate I considering only nuclei larger than r_c. According to their model, an atom which should be added to such a nucleus has to overcome the energy barrier U_i. Adding this barrier to the energy term in formula 2.5 leads to the nucleation rate

$$I = I_0 \cdot e^{-\frac{\Delta G_c + U_i}{k_B T}}. \qquad (2.6)$$

Using the relation between the viscosity η and the energy barrier U_i [Uhlmann:1972]

$$\frac{1}{\eta} \propto e^{-\frac{U_i}{k_B T}}, \qquad (2.7)$$

the nucleation rate can be rewritten as

$$I \propto \frac{1}{\eta} \cdot e^{-\frac{\Delta G_c}{k_B T}}. \qquad (2.8)$$

This formula describes the dependency of the nucleation rate on temperature, neglecting terms which are only linear in temperature. It shows that nucleation is a statistical process with increasing probability at higher temperatures.

Chapter 2

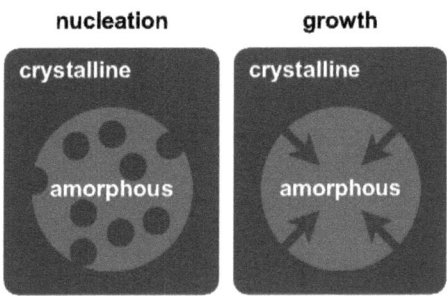

Figure 2.1: Crystallization mechanisms: nucleation and growth. While crystalline nuclei can be formed everywhere in the amorphous volume, the growth mechanism describes an atomic rearrangement from existing crystalline rims.

2.1.2 Crystallization: Growth

In contrast to the nucleation mechanism, which occurs in the whole amorphous volume, the second important mechanism, crystalline growth, describes the change of atomic bonding at the interface between existing amorphous and crystalline phases. The interface energy U_i hampers the atoms at the interface to change their bonds from the amorphous local order to the crystalline pattern, which would minimize the energy of the entire material. With increasing temperature, there is an increasing probability for each atom to arrange itself in the crystalline order of the neighboring atoms and a decreasing probability to change back to the energetically more favorable bonding situation for this single atom. Subtracting those two probabilities, and using equation 2.7, the speed of crystalline growth u can be written as [Uhlmann:1972]

$$u \propto \frac{1}{\eta} \cdot \left(1 - e^{-\frac{-\Delta G}{k_B T}}\right). \tag{2.9}$$

2.1.3 Amorphization: Glass Formation

Two kinds of amorphous phases have been investigated in this work. On one hand, the amorphous phase, obtained from a dc magnetron sputter deposition, and on the other hand, the glassy state, obtained from a melt-quench process in phase-change memory cells.

Sputter deposition (see chapter 3) allows to create large scale samples of amorphous materials. Experiments which investigate characteristics of this phase are presented in chapter 6.

In addition to the amorphous as deposited phase, in phase-change memory cells also the melt-quenched amorphous phase can be investigated. To ensure a homogeneous initial state, the memory cells will be crystallized in an oven before switching experiments are performed. To recreate an amorphous state in cells, the material has to be heated up above the melting temperature. The kind of the subsequent cooling of the material influences the kind of solidification. Cooling down the liquid phase fast enough will avoid the transition to the energetically more favorable crystalline phase by increasing the viscosity of the material fast enough to freeze the liquid state and preserve the disorder of the atomic arrangement.

The ability to create both crystalline and amorphous volumes in phase-change memory cells using electric pulses allows the systematic investigation of the parameters time and temperature, which are characteristic for the transition behavior of each phase-change material (see chapter 4). In addition, there are interesting electronic phenomena which occur, when switching the materials with electrical pulses. The most significant effect is the so called threshold switching effect which will be discussed in the next section.

2.2 Threshold Switching

In 1968, Ovshinsky found a "reversible electrical switching phenomena in disordered structures" [Ovshinsky:1968]. He described a sudden decrease of the dynamic resistance in amorphous semiconductors, if the applied voltage U_a exceeds a threshold voltage U_{th}. Subsequent lowering of the applied voltage does not change the dynamic resistance, until a second characteristic voltage, the minimum holding voltage U_h, is reached. This threshold switching effect has been the topic of investigation for the last decades, and two competing explanations have been established. On one hand, a couple of models describe the threshold effect as a purely electronic effect due to an exponential increase of the numbers of conduction carriers (section 2.2.1). On the other hand, there are models which explain the change of resistivity by a reversible local atomic rearrangement to the crystalline structure (section 2.2.2).

The nomenclature, used to describe the different aspects of this effect, is similar in all established models. The initial state of all experiments and theories is the amorphous phase of a semiconductor, with no electric field applied. This state is called the amorphous OFF state. Application of very low electric fields allows to determine the low field resistivity of the amorphous material. The current-voltage-characteristic of the OFF state is nearly linear, and therefore, the low field resistivity is ohmic (Fig. 2.2). This is the sub-threshold regime of the applied electric field E_a.

When increasing E_a, until it exceeds the threshold field strength E_{th}, the I-V-characteristics becomes highly non linear. The dynamic resistance decreases by some orders of magnitude, and the I-V-curve becomes discontinuous (snap back effect). With this threshold switch, the material is excited to its amorphous ON state. At this point, the models differ from each other. While pure electronic theories assume no change of the local atomic order, and a preservation of the amorphous state, other theories assume a change of the local atomic arrangements. There-

Chapter 2

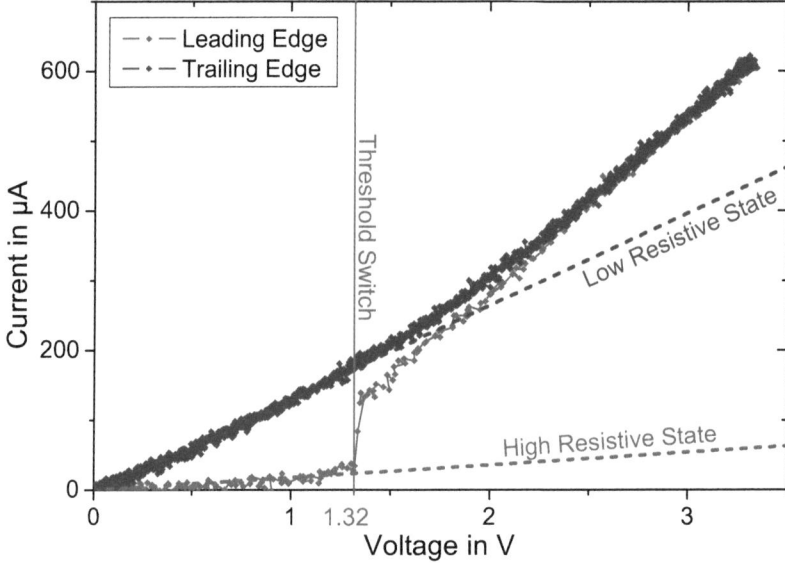

Figure 2.2: Threshold switch in a $Ge_2Sb_2Te_5$ memory cell. The high resistive OFF state has a shallow slope in the I-V-curve (red data points, leading edge of the test pulse). When the threshold voltage U_{th} is reached, the material switches to the low resistive ON state. This state is preserved during the pulse's trailing edge (blue data points).

fore, the nature of the ON state is still topic of discussion, and the time-resolved observation of the transitions from OFF to ON state, and vice versa (chapter 5), has to provide the necessary data to confirm the correct model.

2.2.1 Generation and Recombination of Carriers

At first, the purely electronic models will be discussed. Adler et al. [Adler:1978] published a model which explains the exponential increase of the number of conduction carriers due to impact ionization caused by the accelerated carriers in the amorphous material. This model has been modified by Pirovano et al. [Pirovano:2004c, Redaelli:2008], and numerical simulations have been compared with experimental results.

Ovshinsky supported a purely electronic mechanism primary related to the balance between a strong Shockley Hall Reed (SHR) recombination through trap levels and a generation mechanism driven by both electric field and carrier densities [Adler:1980, Redaelli:2008] (compare Fig. 2.3).

Redaelli describes the avalanche like carrier generation rate, due to impact ionization, by [Redaelli:2008]

$$G = A(n+p) \cdot g(E) \tag{2.10}$$

where n and p are the carrier densities of electrons and holes, A is a constant, and g(E) is a monotonic function. According to Redaelli, the recombination rate can be written as [Redaelli:2008]

$$R^{SHR} = \frac{np - n_{i,\text{eff}}^2}{\tau_h(n + n_l) + \tau_e(p + p_l)}, \tag{2.11}$$

$$\text{with} \quad n_l = n_{i,\text{eff}} \cdot e^{\frac{E_{\text{traps}}}{k_B T}}, \tag{2.12}$$

$$\text{and} \quad p_l = p_{i,\text{eff}} \cdot e^{-\frac{E_{\text{traps}}}{k_B T}}, \tag{2.13}$$

with the carrier lifetimes of electrons τ_e and holes τ_h, and intrinsic carrier densities of electrons $n_{i,\text{eff}}$ and holes $p_{i,\text{eff}}$.

At low applied fields, generation and recombination can be neglected, which explains the ohmic nature of the low field OFF state. With increasing field strength, the generation mechanism increases the carrier number, but a recombination with trap states inhibits their contribution to the conduction process. Increasing E_a further leads to a saturation of the carrier trapping, due to the decreasing number of unoccupied trap states. Hence, the generation process becomes dominant, and the number of free carriers increases exponentially, which causes the threshold switch.

A second model by Ielmini et al. [Ielmini:2008, Lavizzari:2010] explains the threshold switch using a hopping transport model after Poole-Frenkel. In contrast to Pirovano's model, where the trap states impedes the carrier conduction, Ielmini's model assumes the trap states to be fundamentally necessary for the conduction.

Thermally activated carriers can leave the localized trap states, move, and fall back to another trap state in a distance Δz (compare Fig. 2.4). Those carriers can move in the direction of or against the applied field E_a, which leads to a forward or a backward current. This results in an effective current density,

$$I = 2qN_{T,\text{tot}}\frac{\Delta z}{\tau_0} e^{-\frac{E_C - E_F}{k_B T}} \sinh\left(\frac{qE_a \Delta z}{2k_B T}\right), \tag{2.14}$$

where $N_{T,\text{tot}}$ is the effective defect state density above the Fermi level E_F, while E_C is the mobility edge for the conduction band, and τ_0 is the attempt to escape time from a localized

Chapter 2

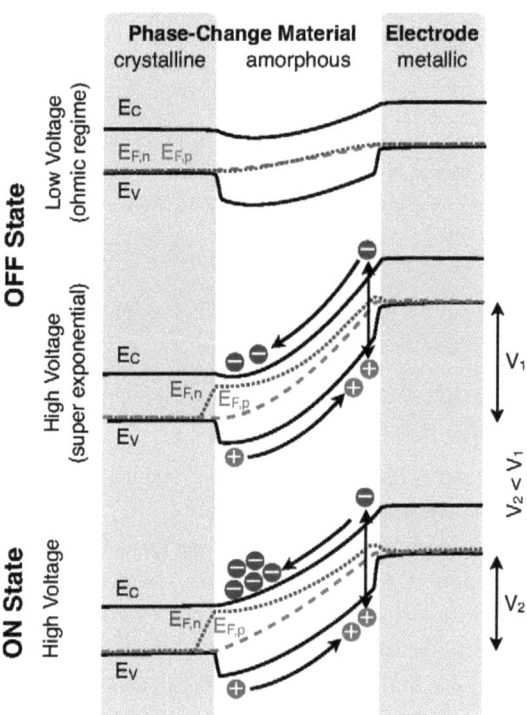

Figure 2.3: Energetic band diagram. Low voltage: the quasi-Fermi levels (electrons, $E_{F,n}$ - holes, $E_{F,p}$) are close to the equilibrium position and the traps are charged. High voltage OFF state: the traps start to be neutralized by generated carriers and the quasi-Fermi levels approach the trap level positions. High voltage ON state: after the filling of most of the traps, the electron quasi-Fermi levels overcome the trap levels moving close to the conduction band. (From reference [Pirovano:2004c, Redaelli:2008])

state [Lavizzari:2010]. An applied electric field E_a reduces the potential barrier for the hopping mechanism, and therefore, leads to an increasing current density following a hyperbolic sine function.

To explain the threshold switching characteristics, the energetic distribution of the trap states has to be taken into account. Lavizarri suggests a simplified model of two discrete trap levels, E_{T1} and E_{T2}, instead of a continuous distribution [Lavizzari:2010]. The generation rate from deep trap states, with a carrier concentration of n_{T1}, to shallow trap states, with n_{T2}, can be modeled by Fowler-Nordheim tunneling:

$$G = \frac{n_{T1}}{\tau_0} e^{-\frac{B_{12}}{E_a}} \quad (2.15)$$

$$\text{and} \quad \frac{د \delta n_{T2}}{dt} = G - \frac{\delta n_{T2}}{\tau_n}, \quad (2.16)$$

where B_{12} is the Fowler-Nordheim coefficient, and τ_n is an effective energy relaxation time for

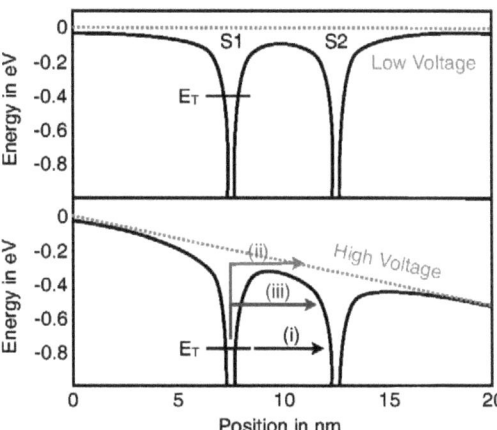

Figure 2.4: Electron potential energy along the minimum path between localized states S1 and S2 for low (top) and high (bottom) applied voltages. Electron transport processes via localized states: (i) tunneling through the energy barrier at E_T, (ii) thermal emission over the energy barrier and (iii) thermally assisted tunneling through the energy barrier at an energy $E > E_T$. (From reference [Ielmini:2008])

electrons [Lavizzari:2010]. Considering these two trap state levels, equation 2.14 has to be modified to sum up the current contributions of both channels,

$$I = \left(2qn_{T1}e^{-\frac{E_C-E_{T1}}{k_BT}} + 2q(n_{T2}+\delta n_{T2})e^{-\frac{E_C-E_{T2}}{k_BT}}\right)\frac{\Delta z}{\tau_0}\sinh\left(\frac{qE_a\Delta z}{2k_BT}\right), \qquad (2.17)$$

where δn_{T2} is the excess carrier concentration at E_{T2}, due to the energy gain mechanism [Lavizzari:2010].

Both models are suitable to use computer simulations to reproduce the published data regarding threshold switching. The threshold switching delay time τ, which has been observed by the aforementioned authors, can be calculated using a numerical approach to solve the equation systems. Unfortunately, both models do not provide an analytical formula for τ, in contrast to the model presented in the next section.

2.2.2 Electric Field Induced Nucleation

A completely different approach has been published by Karpov [Karpov:2008, Karpov:2008b, Nardone:2009], and has been strengthened by Kohary [Kohary:2011]. According to their model, the high electric field induces a local nucleation, which leads to a change in resistivity, respectively to the resistivity contrast of the phases. They assume "a lowering of the free energy of

Chapter 2

the system due to the reduction in the electrostatic energy" [Nardone:2009],

$$W_E = -\frac{\Omega E_a^2 \epsilon}{8\pi n}, \tag{2.18}$$

where ϵ is the dielectric permittivity of the amorphous material, and Ω is the volume of the nucleus. The factor n incorporates the local changes of the applied field due to the exact geometry of the nucleus.

Due to the smaller resistivity in the crystalline nucleus, the voltage drop over the nucleus becomes smaller, and therefore, the field strength in the residual amorphous volume increases. The threshold field E_{th} necessary to grow a nucleus to a conducting filament between the electrodes is given as

$$E_{th} = \frac{1}{\ln(\tau/\tau_0)} \frac{W_0}{k_B T} \sqrt{\gamma \frac{W_0}{\epsilon}}, \tag{2.19}$$

where W_0 is the barrier energy at zero field, and τ_0 is the characteristic vibrational time [Nardone:2009]. The geometric properties of the nucleus and the electrodes are summarized in the factor γ.

The pure threshold switch, without consequent memory switch, can be explained by the subcritical radius of nuclei and filaments, which are only stable in high electric fields, but will decay when the field is switched off. From this theory, Karpov et al. could estimate the threshold switch delay time

$$\tau = \tau_0 \cdot \exp\left(\frac{W_0}{k_B T} \frac{\tilde{E}}{E_a}\right), \quad \text{when} \quad E_a > \tilde{E}. \tag{2.20}$$

Unfortunately, the authors do not provide a physical interpretation of $\tilde{E}$, other than a comment that it corresponds to $W = W_0$ [Karpov:2008].

Despite the different assumptions for both conduction and threshold switching mechanism, all three models can reproduce the experimental data. The differences and uncertainties of the time resolved data, found in literature, are larger than the differences between calculations of the three models. Therefore, it is indispensable to improve the accuracy of time resolved current and voltage measurements to reduce the error of the experimental data below the size of the significant features of the models. This challenge was one of the main tasks of this work, and led to the development of the pulsed electrical tester (PET, chapter 3). The data regarding threshold switching, obtained with this setup, will be presented in chapter 5, where the threshold delay time plays a major role.

3 Experimental Methods

Switching experiments in phase-change memory cells demand a successful interplay between nanometer-sized test devices and gigahertz probe electronics. The dimensions of the test devices have to be of the order of some tens of nanometers for two reasons: to enable high current densities using low voltage sources to melt the phase-change material, and to allow fast cooling rates of ~100 K/ns using the advantageous ratio of active volume to surface of the cooler surrounding material for melt-quenching. The fast crystallization process necessitates electric generators with pulse lengths of a few nanoseconds and gigahertz oscilloscopes to record the current characteristics on a picosecond timescale, while the resistance of the phase-change cell changes some orders of magnitude and complicates the impedance matching. The following sections present one approach to handle these challenges.

3.1 Phase-Change Memory Cells

Two concepts of phase-change memory cells have been established: the mushroom-cell [Lai:2001] and the line-cell [Lankhorst:2005]. Both concepts use the Joule heating of an electric pulse to induce the phase-change in a nanometer-sized volume. The line-cell concept utilizes the resistance of the phase-change material itself to produce the necessary heat, while the mushroom concepts uses a small electrode to generate heat less dependent on the resistance of the phase-change material.

For a material screening, the mushroom design holds some advantages, and this work focuses on this concept. Whereas line-cells demand a nanometer-scaled lithography of the phase-change layer, the mushroom concept can be realized with micrometer-scaled phase-change layer lithography. This allows a simplified deposition and structuring of the phase-change layer on pre-structured samples which provide the nanometer-sized bottom electrode. In addition, the resistance of the active material's crystalline phase, or more precisely of its high-field ON state, plays a minor role during the joule heating process, and the resistance of the bottom electrode becomes more important. Therefore, the mushroom-cells design allows an easier comparison of different phase-change materials regarding pulse heights and induced current flows.

A simplified cross section of the used samples is shown in Fig. 3.1. The phase-change layer is

Figure 3.1: Simplified cross section of a phase-change memory cell in mushroom design. The phase-change layer is sandwiched between the large area top electrode and the small heater, which is connected to the bottom contact pad by a buried tungsten layer.

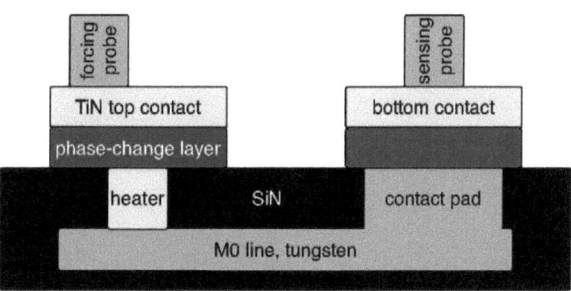

sandwiched between the small bottom electrode, called *heating electrode*, or just *heater*, and the large area top electrode. This leads to an increased current density in the bottom electrode, and in combination with a moderately resistive material, like titanium nitride[1], to a well controllable heat source.

The top electrode is shaped to a L-form (Fig. 3.2) and works also as contact pad, where the forcing probe of the electrical tester is attached directly, while the bottom electrode is connected to a contact pad by a tungsten wire, buried in the dielectric, on which the sensing probe is positioned (compare section 3.2.1). Due to the lithography processing (see section 3.1.1), there is a phase-change layer between the basic metallic M0 line, that connects heating electrode and bottom contact pad, but the large contact area avoids any influence on the switching experiments.

3.1.1 Sample Processing

State-of-the-art sample processing requires high-end lithography tools which themselves require the strictly controlled environment of industrial cleanrooms. To avoid cross contamination of different materials and alloys used in modern semiconductor devices, a candidate material for novel devices has to be investigated regarding these concerns before it can be introduced to the fabrication and characterization lines. These clearance procedures delay and challenge a material screening using novel material compounds with not characterized atomic species. Therefore, university and industry can benefit from each other by combining their strengths and compensating their weaknesses. Regarding phase-change memory research, a university group can provide a broad material repertoire, in terms of deposition and characterization, while the industry can provide reliable nanometer-lithography methods, applied on established materials. This leads to the idea of a research project with the german chip manufacturer Qimonda: test wafer with heating electrodes of 60 nm diameter, processed in the chip factory

---
[1] Resistivity of TiN ≈ 200 µΩcm.

Experimental Methods

Figure 3.2: Qimonda samples. The 300 mm wafer (background) was cleaved into pieces with only one test chip (middle). Each test chip provides 120 test cells (front), consistent of a L-shaped top electrode and a square-shaped bottom contact pad, which is connected via a buried tungsten line with the 60 nm diameter heating electrode.

in Dresden, should be finished in the university labs of Aachen, using µm-lithography to deposit and structure various phase-change materials.

Qimonda provided 300 mm test wafer with around 340 test chips with 120 pre-structured memory cells on each chip (Fig. 3.2). Each memory cell consists of two contact pads and a heating electrode, which is connected to one of the contact pads by a buried tungsten line. The contact pads (located on the very right side in Fig. 3.2) are tungsten squares of 40 µm times 40 µm. The bottom pad (right) is connected to the M0 line, a 300 nm broad tungsten line that is buried in the surrounding SiN dielectric. At the end of this 60 µm long line is the heating electrode (Fig. 3.4). The heating electrode was created by etching a 60 nm diameter hole in the dielectric, which was filled with titanium nitride. Afterwards, the surface was planarized with the help of chemical-mechanical polishing (Fig. 3.1).

Chapter 3

On this planarized surface, the phase-change layer and the top contact were deposited and structured in Aachen (Fig. 3.3). Because phase-change materials are much weaker than the titanium nitride of the top layer, etching is not suitable to form the top contact, and therefore, a lift-off process was chosen. For the lithography process, a single chip was cleaved from the pre-structured wafer and was covered with photoresist. Afterwards, a contact mask aligner was used to expose the L-shaped area for the top electrode and the square bottom contact pad. The exposed photo resist was removed, and the phase-change layer and the titanium nitride electrode material were deposited on the sample to fill the exposed areas. Using acetone the excessive material on the non-exposed areas was removed together with the remaining photoresist.

All lithographic steps were performed in the cleanroom of the II. Institute of Physics using the photo resist ARU 4040, a spincoater (5000 rpm), a contact mask aligner (20 s) and a chemical developer (40 s). The contact mask was designed in cooperation with Qimonda. To maximize the transparent areas on the contact mask, a negative exposure was chosen for the lithographic process. Hence, only the L-shaped top contact and the small square of the bottom contact (Fig. 3.2) inhibit the sight through the glass plate of the contact mask. This allows an easy and precise mask alignment on the pre-structured sample within ±3 µm.

After the lithographic steps, the sample was etched in hydrofluoric acid (HF-dip in 1 vol% for 120 s) to clean the surface of the heating electrode from oxides. Afterwards it was mounted in the deposition chamber immediately. To ensure an evaporation of the hydrofluoric acid residuals, the chamber was evacuated for 3 hours leading to a basic pressure of $1.2 \cdot 10^{-6}$ mbar. Subsequently, the deposition process was initiated by inflating 20 sccm argon to generate the process atmosphere of $3.5 \cdot 10^{-3}$ mbar. In this atmosphere, a direct current plasma was ignited using a power controlled voltage supply set to 20 W output. The phase-change material was sputtered from a stoichiometric compound target mounted on a magnetron cathode. Subsequently to the pre sputter process against a blind, a 20 nm thick film was deposited on the sample.

Without breaking the vacuum, the top electrode was deposited on top of the phase-change layer. To decrease the stress between phase-change layer and titanium nitride, and to enhance the adhesion of these layers, a 4 nm thick titanium layer was deposited from a titanium target in a 60 W dc magnetron sputtering process (8 sccm argon flow). The 40 nm thick inert top electrode was sputtered afterwards from the same titanium target in a reactive process, adding 1.8 sccm nitrogen to the process gas, and increasing the power to 120 W.

In an ultrasonic bath (2 min) of acetone, the lift-off process was performed. Subsequently, the sample was annealed at 200°C for 10 min to crystallize the phase-change material (250°C for $Ge_{15}Sb_{85}$) by heating up and cooling down with 10°C/min.

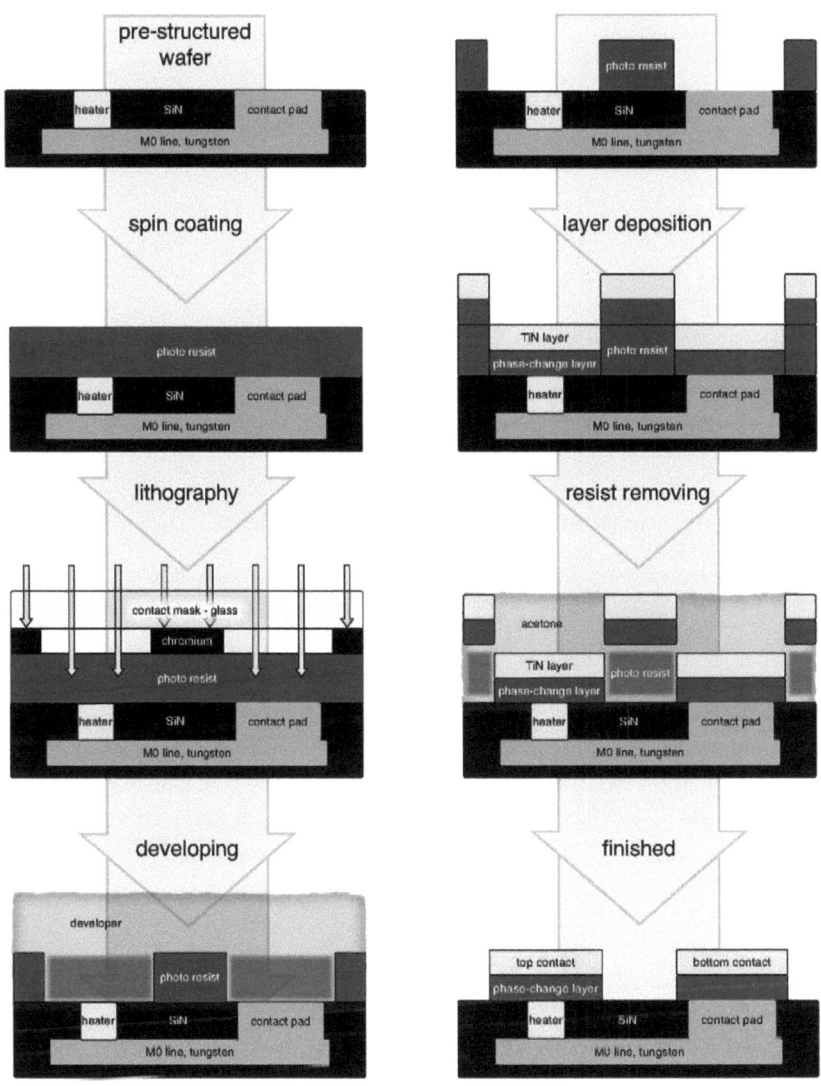

Figure 3.3: Lithography and lift-off process. Lithography: a photo resist is spin-coated on the pre-structured wafer; the areas over heater and contact pads are exposed using a contact mask aligner; the exposed photo resist is removed by a chemical developer. Lift-off process: a phase-change layer and a titanium nitride layer are deposited on the structured photoresist; dissolving the photoresist with acetone removes the material which has been deposited on top of the photo resist.

Chapter 3

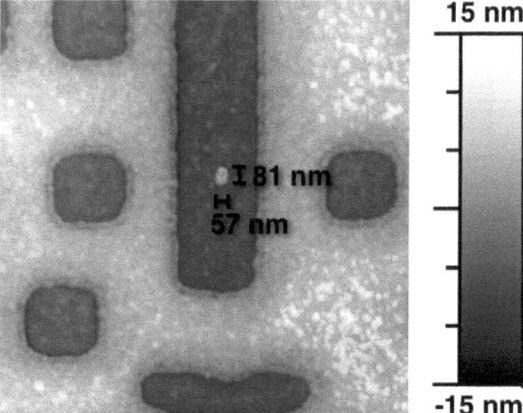

Figure 3.4: AFM picture of bottom electrode after HF-Dip. Dark colors indicate lower levels and show the position of M0 line and other tungsten support structures. The heater protrudes about 8 nm from the surrounding SiN layer.

3.1.2 Challenges with State-of-the-Art Nanodevices

During the progress of the project, the pre-structured wafer were stored in argon atmosphere, but this could not prevent completely that the oxide layer of the heating electrode grew. Therefore, the hydrofluoric etching had to be adjusted. A second etching step (1 vol%, 120 s) was performed, previous to the spincoating of the photoresist, to avoid underetching and displacement of the photoresist after the development step.

The heating electrode is a very fragile structure, and the etching steps, to clean its surface, change its shape in the dielectric surrounding (Fig. 3.4). In the beginning of the project, the etching times could be decreased to a minimal time, to avoid significant changes, and differences between the heating electrode's shape on one chip, and even to chips that were processed later (see chapter 4.3).

3.2 PET: Pulsed Electrical Tester

Creating phase-change memory cells is one precondition to perform switching experiments, the second is building a setup that can generate the switching pulses, record the events, and provide an interface to connect the sample and the electronic devices. The Pulsed Electrical Tester (PET) fulfills these requirements. Its focus is on high time resolution of the current flow induced by an applied voltage pulse.

There are numerous challenges developing such a setup. Some are correlated to the interplay of the material's properties, others result from the nature of the effects which shall be inves-

Experimental Methods

Figure 3.5: Picture of the PET contact board. Two probing needles connect the board through a 3 cm diameter hole with the 40 µm small contact pads on a test chip. The forcing probe (left) is connected to the generators, while the sensing probe is connected to the shunt resistor.

tigated. On the one hand, there is the large resistivity contrast between the amorphous and the crystalline phases that leads to a broad spectrum of memory cell resistances, spanning over orders of magnitude.

On the other hand, there is the high velocity of the dynamics in phase-change materials. Crystallization takes place within some nanoseconds, and changes of the electronic properties can be even faster. Therefore, it is quite challenging to design an impedance matching network that allows to apply gigahertz voltage pulses on a test device that changes its resistance some orders of magnitude during the experiment, and sometimes within one nanosecond.

Another topic are parasitic capacitances and inductances. Changes of the resistance of the memory cell during a pulse will lead to changes in the electric potential, and therefore, to loading or unloading of the parasitics which leads to disturbing signals, or even to current peaks which could destroy the sample.

The main idea, behind building the PET, was to reach the physical and electronic limitations in terms of maximal bandwidth and best signal-to-noise ratio to investigate the electronic switching in phase-change materials using time resolved current measurement. Therefore, a

Chapter 3

custom-made contact board (Fig. 3.5) was designed to connect the electronic devices and a single memory cell on a test chip. The maximum frequency f_{limit} which can be achieved in a electronic circuit depends on the time constants τ of its components.

$$f_{limit} = \frac{1}{2\pi\tau} \tag{3.1}$$

Each time constant τ describes the necessary time to load or unload each capacitor C in the circuit. Depending on the resistance R which limits the loading current of the capacitor, the time constant τ can be calculated.

$$\tau = R \cdot C \tag{3.2}$$

Therefore, all capacities in the circuit have to be reduced to minimum values and all capacitors have to be connected with low resistances. The signal height, and therefore, the signal-to-noise ratio of the time resolved current measurements depends on the resistance ratio of the device under test R_{DUT} and the shunt resistor R_S.

$$V_{applied} = V_{DUT} + V_{R_S} \tag{3.3}$$

$$\frac{V_{R_S}}{V_{DUT}} = \frac{R_S}{R_{DUT}} \tag{3.4}$$

$$\Rightarrow V_{R_S} = \frac{R_S}{R_{DUT} + R_S} \cdot V_{applied} \tag{3.5}$$

Hence, the accessible frequency limit of the setup is strongly correlated with the desired signal quality. As soon as all capacities are reduced to the physical limitations (e.g. minimized contact pads on the sample) the frequency limit can not be improved without loosing signal quality.

Therefore, the first goal during the construction of the PET was to minimize all capacities in the setup. Afterwards, the system was tuned to the most promising compromise between signal speed and quality. Due to its design, the PET can be readjusted easily to change this compromise to higher frequencies or to better signal-to-noise ratio.

The components can be divided into three fields of function: fast time resolved measurements, precise resistance determination, and supply systems.

Time resolved measurements. Time resolved measurements are performed using a voltage source and a fast oscilloscope. Two different voltage sources are employed. On one hand, an arbitrary waveform generator (AWG)[2] allows experiments up to 2.0 V using pulses of arbitrary shape. On the other hand, a pulse generator[3] allows experiments up to 10 V, but can only

---
[2]Tabor ww1281A, max. 2.0 V, min. 800 ps pulse edges.
[3]Hewlett Packard 81110A, single pulse generator, max. 10 V, min. 1 ns pulse plateau length, min.

Experimental Methods

provide trapezoidal pulses. The applied voltage and the induced current are recorded with a digital oscilloscope. Some earlier experiments have been performed with a Tektronix TDS 620B, later the scope has been replaced by a faster LeCroy SDA 13000. With this change, the time resolution of the recorded signals have been improved from 400 ps between two data points to 50 ps. Furthermore, the old scope had only two recording channels while the new scope provides four channels. This allows to record the current signal with two differently amplifying channels simultaneously. The low voltage part of high voltage signals can be recorded in this additional channel and improves the data quality significantly.

Resistance measurement. The cell resistance is determined using a sine wave generator (Agilent 33120A) and a lock-in amplifier (EG&G Princeton Applied Research 5210).

Supply systems. At last, supply devices are necessary to provide an inert atmosphere, a controlled sample temperature, and possibilities to contact the 40 µm-pads on the chips.

3.2.1 Setup Description

The PET[4] is designed to handle memory cells with resistances from $\sim$500 Ω to $\sim$80 MΩ, but can be adjusted to work with smaller or larger resistances by changing the shunt resistor. Obviously, the memory cell can not be matched easily to 50 Ω-devices. Therefore, pulse generator and memory cell (device under test, DUT) are connected via an impedance matching resistor network, as shown in Fig. 3.6 and in Fig. 3.5. A probing needle[5] (forcing needle) connects the resistor network with the top contact pad of the memory cell.

The impedance matching network allows the feedback of the applied signal to an oscilloscope channel. This signal is weakened by a factor of $\sim$6.6[6], while the signal of the pulse generator passes the network unaltered towards the DUT.

A second probe (sensing needle) connects the bottom contact pad of the memory cell (Fig. 3.1) with the contact board. This probe is linked with the shunt resistor (connected to ground) and the circuit for measurements of the current. The applied voltage is divided between DUT and shunt resistor which allows the measurement of the current through the DUT by measuring the voltage drop over the shunt resistor. To reduce parasitic capacitances the probing needles have to be as short as possible [7], and the shunt resistor should be small, too (both in geometry and in resistance). One of the major challenges in designing the PET was to ensure low parasitic

---

[2] ns pulse edge length.
[4] Details regarding the electronics, especially the contact board in [Schlockermann:diss].
[5] Made from copper-beryllium to allow attachment on the contact board by soldering.
[6] The exact value of this factor depends on the used contact board. Four slightly different boards can be mounted in the setup to meet the requirements of different types of memory cells and different aims of experiments.
[7] At least shorter than half the wavelength of the shortest applied pulse.

Chapter 3

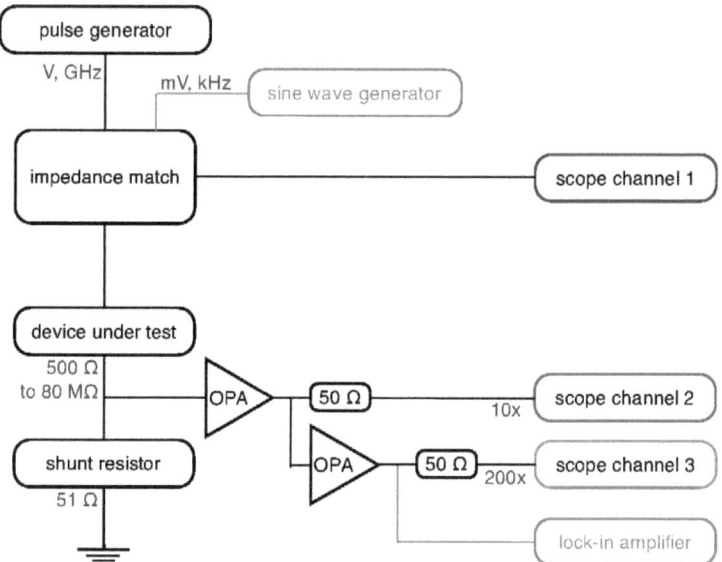

Figure 3.6: PET schematic. Combination of two measuring circuits. Nanosecond pulses can be created and recorded using pulse generator and oscilloscope with gigahertz bandwidth. Device resistance after each pulse can be determined using sine wave generator and lock-in amplifier with kilohertz bandwidth.

capacitances, but enable good current signals.

This has been realized by using a 51 Ω-shunt resistor, and amplifying the small voltage signal over this resistor by two operational amplifiers (OPA). The chosen type of OPA (Texas Instruments OPA847) excels at very good signal-to-noise ratio, a good bandwidth (300 MHz) and a very smooth filter that allows acceptable amplification even for 1 GHz-signals. A noticeable constraint is the optimal amplification factor (20x) of this OPA and the correlated output signal height limited to ∼3 V. Therefore, the input signal should not exceed ∼0.15 V, which can be achieved by adjusting the parameters of memory cell, shunt resistor, and applied voltage pulses.

Memory cells, as presented in chapter 3.1, have a minimal resistance of ∼1 kΩ[8], and maximum voltages of 3 V are necessary to perform all kind of switching experiments. Therefore, the voltage drop over the shunt resistor of 51 Ω will not exceed ∼0.15 V.

Time resolved measurements at higher ohmic cell states require an additional amplification

---

[8]The value of a shorted cell, where only heating electrode and the M0-line provide resistance, comparable with a cell in its crystalline or amorphous ON state.

Experimental Methods

step to boost the signal height above typical noise levels of oscilloscopes. Hence, the output signal of the first OPA is split, and connected both to an oscilloscope and a second identical OPA, where the signal is amplified again, by the factor of 20, to an overall amplification of 400, and connected to an additional oscilloscope channel. All channels of the oscilloscope are internally terminated with $50\,\Omega$, and to ensure impedance matching on the OPA's side, a $50\,\Omega$-resistor is attached in series directly after the signal splitting point. Therefore, only half the voltage is recorded with the scope, which leads to an amplification factor 10 of the signal at channel 2, and a factor 200 for the signal at channel 3.

A precise determination of high resistances, like those of memory cells in the amorphous state, is hardly possible with that configuration. Hence, a second circuit for resistance measurements is added to complement the fast circuit for time resolved measurements. A lock-in amplifier, attached to the 400x output of the second OPA, is synchronized with a sine wave generator, attached to the impedance resistor network, to add continuously a mV-signal to the DUT. This low voltage signal does not disturb the time resolved measurements, but using the lock-in technique, cell resistances up to $80\,M\Omega$ can be determined. Considering the high temperature dependance of amorphous phase-change materials, a precise reliable resistance measurement demands for accurate control of the sample's temperature. Therefore, the sample holder consists of a copper plate and two Peltier-elements to stabilize its temperature, using a water cooled aluminum block as heat sink. Temperatures from -40°C to 120°C can be achieved with a stability of ± 0.01°C.

Using simultaneously two different measurement techniques, connected by an impedance matching resistor network, has two advantages: At first, it allows the specialization of both circuits to their aims: one for fast and time resolved signals, another for low voltage and high ohmic resistance read outs. In addition, this method minimizes the parasitic effects, due to capacities and inductivities, which lead to unwanted voltage and current signals caused by the sudden changes of the resistance of the device under test.

3.2.2 Software and Data Analysis

Already a single pulse experiment demands for a complex interplay of the electronic devices, and produces a high amount of data that has to be analyzed. Two software packages were written allowing the performance and interpretation of elaborate switching experiments which consist of thousands of recorded pulses.

The hardware of the PET is controlled by a LabView program that manages the interplay and timing of the electronic and the supply devices. Its modular style allows an easy implementation of new devices and changes in the chain of commands. While all basic routines are programmed in that LabView code, and can be operated by a few mouse clicks, the design of a multi pulse experiment has to be written in a custom-made script language that allows the usage of variables and for-loops to change pulse parameters systematically. Therefore, the programming of an arbitrary switching experiment necessitates not more than some lines of written code.

Each pulse event creates two data files. One contains the oscilloscope's traces, the second contains all metadata, like set parameters, single value measurements, and setup parameters. These data files are numbered and stored in a folder system, representing the investigated material, the used piece of wafer, and the cell number.

The analysis software is written in MatLab. It uses the same folder system as the LabView control software, to load the raw data and to store the results in a corresponding results folder system. Therefore, an accidental mixing of data from experiments on different chips or materials is excluded. Another benefit of this fix data storage system is the possibility to program the analysis algorithms as generally usable modules, avoiding unnecessary work, when changing the investigated material or single modules of the experiment's work flow.

A special computer with 16 GByte memory enables comparison of features in hundreds of recorded pulses with thousands of data points in each trace. In principle, two kinds of result plots are essential. The one kind contains all the time resolved data (e.g. Fig. 3.7) and shows the development of voltage, current, or related parameters during a single pulse. Traces of related pulses can be plotted in common axes to allow time related comparison of different pulse set parameters. However, those plots contain a wealth of information, and are hard to interpret. Therefore, a second type (e.g. Fig. 4.3) is necessary which focuses on selected parameters, extracted from the time resolved data, and allows comparison of single aspects.

3.2.3 Calibration and Testing

Reliable results, obtained from switching experiments in phase-change memory cells, require an extensive calibration and testing of the setup, the software, and the samples.

Parameters of the contact board, like amplification factors and signal transmission through cables and connectors, are determined using both precise dc-equipment[9] and a network analyzer[10]. Using non-active memory cells with titanium (short sample) or a dielectric (open sample) instead of a phase-change layer, both extreme signal ways through a memory cell can be tested. With this test cells, the parasitic capacitance ($\sim$100 fF) of probes and samples are determined and considered in the data analysis[11].

The signal quality of both pulse generators is displayed in Fig. 3.7. The shortest possible pulses of both sources are plotted for comparison. While the pulse generator's minimal pulse

---

[9]Keithley 236 sourcemeter.
[10]Rhode & Schwarz ZVL Network Analyzer 3 GHz.
[11]Profound details concerning the parasitic capacitances are reported in [Wimmer:2010].

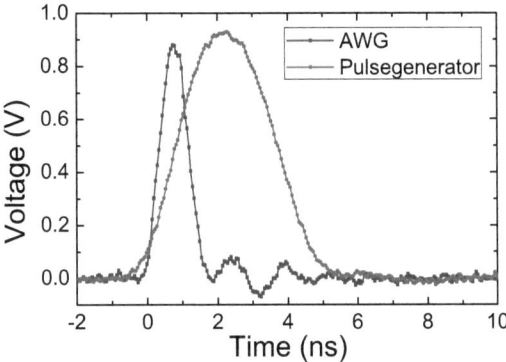

Figure 3.7: Time resolved measurement of the shortest possible voltage pulses that can be generated with the arbitrary waveform generator (AWG) and the pulse generator. The full width half maximum (FWHM) of the AWG is with ~1 ns much shorter than the FWHM of the pulse generator ~3 ns.

edge length is 2 ns the AWG can create 800 ps edges. Hence, the AWG's shortest pulse has a full width at half maximum of ~1 ns, while the pulse generator's is about 3 ns.

Figure 3.7 displays the time resolved voltage measurement on a test cell using the contact board. The time resolved current measurement of cells does not provide the same signal quality, because of additional response times and slew rates due to the OPAs in the current recording path. This difference in important for the interpretation of the experimental results in the following chapters: On one hand it is possible to apply very short and clean voltage signals on a test cells. On the other hand the time resolved current measurements can be affected by additional setup components. Especially signal components above 500 MHz (pulse features below 2 ns) can be altered and damped. In extreme cases like threshold switching experiments with high and steep pulses, the current signal can be suppressed completely.

3.2.4 Pulse Parameter Definition

To simplify the specifications of the set parameters of different pulses, the following notation will be used in the next chapters. The standard pulse, used in the most experiments, are trapezoidal, and therefore, need four parameters to be defined: pulse height (ph), leading edge (le), plateau length (pl), and trailing edge (te). The pulse edge length is defined as the time which is necessary to rise the output voltage of the generator from 10 % to 90 % of the pulse height. Hence, the pulse plateau length describes the time for which the output voltage of the generator is higher than 90 % of its maximum set value. An easy and short way to notate these

Chapter 3

parameters is: ph, le/pl/te.

For example, 1.8 V, 19/128/19 ns will describe a trapezoidal pulse with a pulse height of 1.8 V, a plateau length of 128 ns, and with leading and trailing edge lengths of 19 ns each. The term *pulse length* will be used to describe the effective pulse length of pulses with the shortest possible edges. In this case, pulse length is equal to the plateau length.

3.3 Sheet Resistance Measurements - van-der-Pauw Setup

Switching experiments in memory cells are a powerful tool to investigate the kinetics and electronic effects during the phase transitions on short timescales. They can be complemented with high sensitive electric measurements on long timescales.

Temperature dependent resistance measurements on unstructured thin films provide insight in both crystallization temperatures, thermal activated carrier transport, and allow investigation of the drift mechanism in amorphous phase-change materials (chapter 6). These properties could be investigated using memory cells, but the highly demanding production process would hinder a fast material screening, and the two-point measurement technique is not optimized for recording small changes in resistance. Therefore, amorphous as deposited phase-change layers on insulating substrates and the van-der-Pauw method for sheet resistance measurements [Pauw:1958] were chosen for this task.

The resistance of amorphous phase-change films can reach some hundreds of Gigaohms, and the activation energy for conduction will change the resistance significantly with slightest fluctuations of the sample's temperature. Hence, both a sensitive measurement system and a high-precision temperature control has been built, and will be presented in the following sections.

3.3.1 Setup Description

The van-der-Pauw setup (vdP) is a combination of a contact heater device and a four-point probe system. Probe heads and contact heater (Fig. 3.8) are placed in a vacuum chamber to allow a good control of the ambient atmosphere (argon 6.0) and pressure (5 mbar to 1080 mbar). Guarded triaxial cables connect the probe heads with a custom-made switch matrix[12] to allow the source and sense alternation for the van-der-Pauw measurement technique using a Keithley 236 sourcemeter and an Agilent 34401A voltmeter.

The standard sample is a phase-change layer[13] on a glass substrate[14] with chromium or

---
[12] Profound details regarding the switch matrix can be found in [Schlockermann:diss].
[13] Dimensions: 1 cm×1 cm of 100 nm thickness; deposited by dc magnetron sputtering (sec. 3.1.1).
[14] Dimensions: 2 cm×2 cm of 0.5 mm thickness.

Experimental Methods

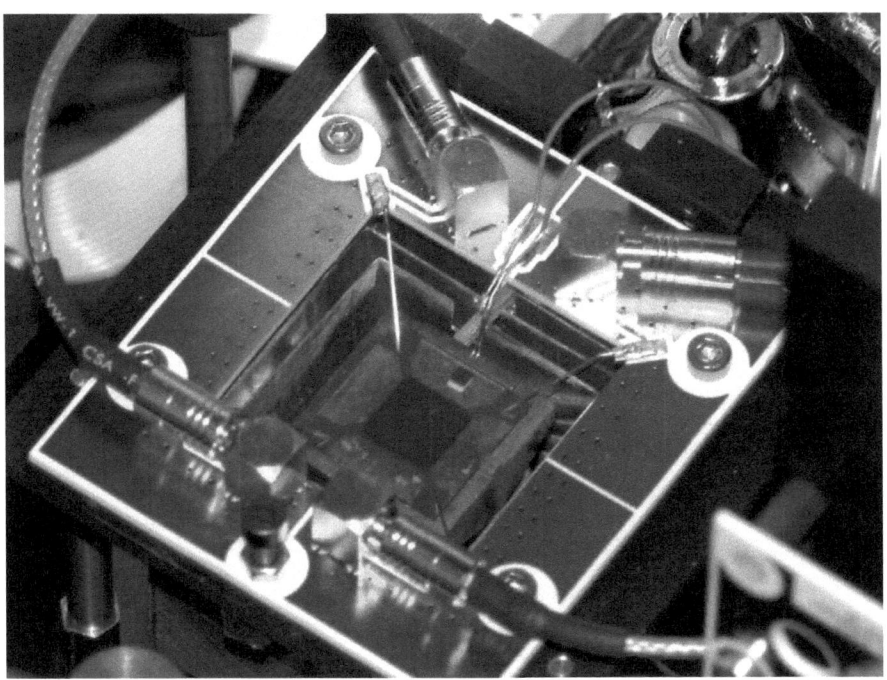

Figure 3.8: Picture of the van-der-Pauw setup. Square shaped phase-change film and chromium contacts on a glass substrate. The probing needles press the sample against the contact heater.

aluminum[15] contact pads, which ensure electrical contact through a not-oxidized interface with the phase-change layer as shown in Fig. 3.9. Both to connect the phase-change film with the electronics and to press the sample towards the contact heater, four probe needles have to be attached to the chromium contact pads.

A simplified cross section of sample and contact heater is shown in Fig. 3.10. The sample holder is custom-made from copper with cavities for a temperature sensor[16] and four electric heating cartridges. A copper tube coil is attached to this copper block to allow fast cooling using compressed air. Via a control box the heater is connected with the computer.

Also chamber atmosphere, electronic devices, and switch matrix are controlled by the computer using a LabView software adapted from the PET. Focus of this software is the precise timing of temperature read out and heater control. The developed script language from the

---

[15] Deposited by dc magnetron sputtering process comparable to process for titanium (sec. 3.1.1).
[16] Precise platinum resistor: PT 100.

Chapter 3

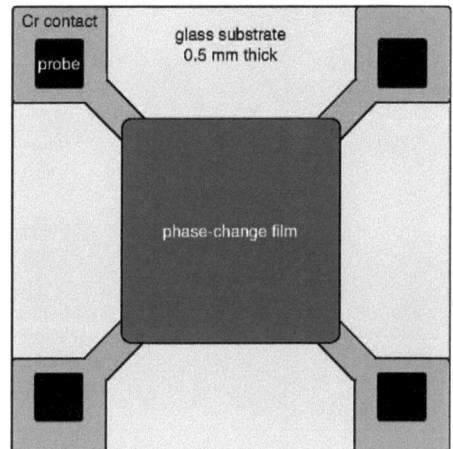

Figure 3.9: Sample geometrie for van-der-Pauw setup. Square shaped phase-change film and chromium contacts on glass substrate. The probing needles connect the sample to the electronic measurement devices, and press the sample against the contact heater.

PET software (section 3.2.2) was adapted, too, and allows an individual design of heating and cooling ramps and annealing times for elaborate experiments.

Calibration experiments[17] have shown that the temperature stability of the copper block is ±0.01°C, and that heating and cooling rates up to 180°C/min are possible. Using additional temperature sensors, attached to standard glass substrates, the surface temperature was measured during different heating/cooling procedures and for different annealing temperatures. Those experiments revealed a small temperature mismatch between the surface of the sample and the copper block. This is considered in a calibration table for the heater control, and leads to a surface temperature accuracy of ±2°C, while temperature fluctuations are smaller than 0.01°C.

Depending on the resistance of the sample, a sheet resistance measurement requires different integration times. Resistances up to $1\,M\Omega$ can be determined within some seconds, while for higher ohmic samples (up to $500\,G\Omega$) the capacitance of sample and setup necessitates integration times up to 1 minute. The software uses a change-of-deviation method to check if the elapsed integration time was sufficient and optimizes the time for each measurement point without losing data precision.

In summary, the experimental tools to investigate phase-change materials in memory cells have been presented in this chapter. Besides the description of a possible way to create state-of-the-art memory cells in mushroom design, the basic working principles of two custom made setups have been explained. The development of both setups, described in this chapter, was

---

[17] Further details can be found in the diploma thesis of Rüdiger Schmidt [Schmidt:2010].

Experimental Methods

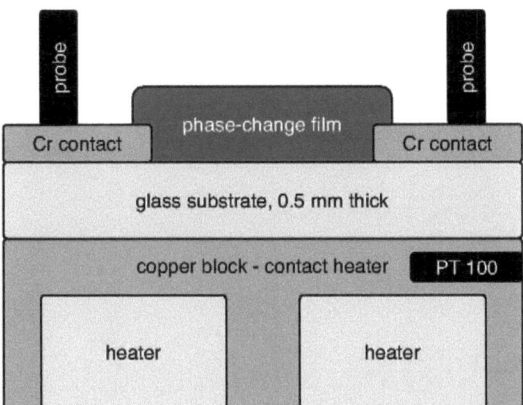

Figure 3.10: Schematic of the van-der-Pauw setup. Square shaped phase-change film and chromium contacts on glass substrate. The probing needles press the sample against the contact heater, which uses four heating cartridges and a copper tube coil for precise temperature regulations. The copper block temperature is measured using a PT 100 platinum resistor.

mandatory to enable the experiments which are presented in the next chapters.

4 Phase Transition in Memory Devices

Driving force of all efforts described in this thesis was meeting the challenge to investigate and to understand the electronic switching mechanisms in phase-change materials. To achieve that goal, test cells (chapter 3.1) were created and a setup was built (chapter 3.2). The next step was to developed measurement routines which allow the systematic investigation of the switching behavior of different phase-change materials.

The first part of this chapter will describe the different pulse patterns which are necessary to switch memory cells between well defined states. Later, the results of these switching experiments on different phase-change materials will be discussed. This chapter will focus on the low-voltage state of the cells in dependence on different shapes and strengths of applied test pulses. The high-voltage states and the phenomena during the electronic switching, like threshold switching and threshold recovery time, are topic of the next chapter.

A powerful method to visualize the effect of different pulse parameters on the cell resistance is a refined power-time-effect (PTE) diagram. PTE diagrams are commonly used in optical phase-change switching experiments to summarize the change of the material's reflectance due to applied laser pulses of different intensity and length. For electronic switching experiments, it is reasonable to change from pulse power to applied voltage or induced current. Therefore, two diagrams will be used in this chapter: current-time-resistance (CTR) and voltage-time-resistance (VTR) diagrams, where the low-voltage cell resistance will be used as the investigated effect. To avoid confusion, both diagrams will be referred as PTE diagrams, as this name has been established for both kinds of diagrams.

4.1 Initialization Procedures in Mushroom Cells

Despite careful and homogeneous sample preparations, the test cells show slight variations. This can have an effect on the switching behavior and on the cell resistances. Before a switching experiment can be performed on a new phase-change cell, some initialization procedures have to be applied to ensure comparable initial states (section 4.1.1).

Another important aspect for reliable experimental data is the handling of the cell's history.

Chapter 4

After SET or RESET operations[1], there can be residual amorphous volumes within the switchable volume of a cell. Those volumes can have a significant influence on the switching behavior. Section 4.1.2 presents some techniques to eliminate the cells history and recreate a well defined initial state on which different pulse parameters should be tested.

4.1.1 Cell Training

The most common effect is a major change of the switching current after a strong pulse was applied. To avoid those characteristic changes of a cell during an experiment, it is useful to apply some pulses in advance, which are higher and longer than the strongest pulse during the experiment. But very long (some µs) and high pulses (>1.6 V) can damage the cell, and therefore, have to be applied with care. Thus, a method was required to prepare the cell for very strong pulses and to minimize changes of its characteristics during experiments.

A pattern of short (<200 ns) pulses with increasing height (up to 2.2 V) was developed. Three pulse types were used in alternation: 1) a long edge pulse with rising pulse height, 2) a long plateau pulse with constant pulse height, and 3) a short pulse with rising pulse height (see Fig. 4.1). The rising pulse height increases the induced heat in the cell and increases the size of the molten volume above the heater. The pulse with the constant height secures the crystallization of material very close to the heater. In combination with the long edge pulse, the crystallized volume of the cell will increase with every iteration, while the short pulse will amorphize the same volume afterwards. Therefore, the phase-change material in the cell will be crystallized and amorphized consequently, and the atoms of the active volume will be mixed. Due to the increasing of the pulse height, this mixing will be expanded to a larger volume with every iteration.

Cells which were initialized using this pattern showed very reliable and reproducible results during the experiment, and repetitions of the experiments reproduce those results within the errors.

4.1.2 Recreate Initial State

Testing the influence of different pulse parameters on various cell states leads unavoidably to incomplete SET and RESET states and residual amorphous clusters within the switchable volume of the cell. Hence, it is important to delete the cells history before the next test pulse is applied. As shown by Lacaita et al. [Lacaita:2004], there is a broad variety of residual amorphous configurations for incomplete SET states. To recrystallize the programmable volume of the cell, it is essential to heat up the whole volume long enough, and to cool it down using a

---

[1]See chapter 1.2.

Phase Transition in Memory Devices

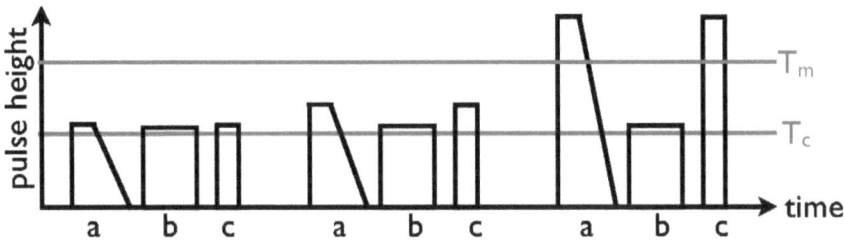

Figure 4.1: Pulse pattern for cell training. Three pulses have been repeated with increasing heights of pulses a) and c). a) Pulse with long trailing edge. b) Pulse with long plateau length. c) Short pulse.

shallow voltage ramp. This could be realized with a pulse of the maximum height used in the experiment and a trailing edge of some microseconds. But such a pulse will destroy the cell after some hundreds of repetitions, and is therefore not suitable for elaborate experiments with thousands of parameter variations.

A more gentle treatment of the cell can be realized by using two pulses in combination to erase all amorphous residuals. A first pulse with the maximum height, used in the experiment, but with an very short plateau length of some nanoseconds and a trailing edge of around 200 ns. This pulse will heat and crystallize the outer part of the switchable volume, but cause a melt-quenching of the inner part next to the heater. A second pulse with a moderate pulse height, but a long plateau, can recrystallize that inner part without melting any volume within the cell. Using this pulse pattern, the cell can be recrystallized reliably without damaging the cell, and it secures the absence of any amorphous residuals.

4.2 Nanosecond Switching in GeTe Memory Cells

A large variety of phase-change materials has been investigated during the last decades. Most materials which have been proposed as candidates for phase-change memory were already implemented successfully in optical data storage. $Ge_2Sb_2Te_5$ and AgIn-doped Sb_2Te can be found in rewrite-able CDs and DVDs, and the results of their electronic switching experiments are shown in section 4.3 and 4.4.

In this section, the results of GeTe will be presented; a material, which has left the focus of investigations in the late 1980's, because two properties of GeTe complicate the integration in industrial applications. At first, there is a a strong dependency of crystallization speed and crystallization temperature of Ge_xTe_{1-x} on the exact value of x, as reported by Chen et al.

Chapter 4

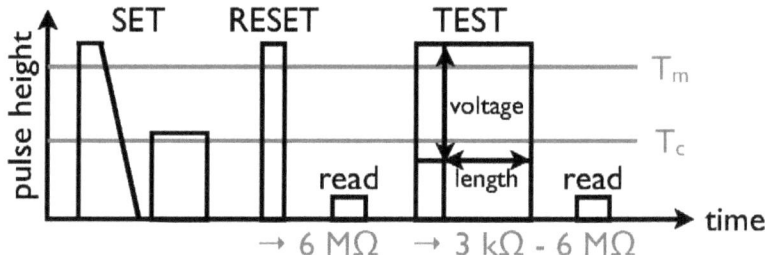

Figure 4.2: Pulse pattern. To ensure a uniform and reproducible initial state, each test pulse was preceded by both a full SET (3 kΩ) and a subsequent RESET (6 MΩ).

[Chen:1986]. Secondly, there is a significant dependency of x on the exact parameters of the sputtering process. The combination of these two properties results in a large spread of nucleation speeds. Therefore, this material became less interesting for the industrial applications, although Chen et al. already reported recrystallization speeds in the range of 30 ns.

The following section will show results and explanations why GeTe is one of the most promising candidates for a phase-change memory. Due to its switching time of 1 nanosecond [Bruns:2009], it can compete with the write and read times of a state-of-the-art DRAM.

4.2.1 Experimental Parameters

As crystallization is the time critical transition for a high speed memory device, an experiment was designed to determine the minimum pulse length necessary for a complete SET of the cell. The influence of both pulse length and pulse height should be investigated, which leads to a two dimensional parameter scan (Fig. 4.2). Each iteration step consists of four single pulses and two read out measurements. At first, a secure SET was performed using a two pulse method[2] (see section 4.1.2) that leads to a cell resistance of 3 kΩ. Afterwards, the cell was switched to the RESET state using a short amorphization pulse[3]. The resulting resistance of the cell was determined carefully using the lock-in technique (Chapter 3.2.1).

Subsequently, the test pulse was applied, followed by another resistance measurement. For the first part of the experiment presented, the test pulse height was increased starting from 0.6 V up to 2.0 V with an increment of 0.1 V. When the maximum value was reached, the pulse height was set back to the start value for the next iteration, and the pulse plateau length was

---
[2]Pre-crystallization: 2.0 V, 19/20/190 ns; Init-crystallization: 1.0 V, 19/128/19 ns
[3]RESET pulse: 2.0 V, 19/16/2 ns.

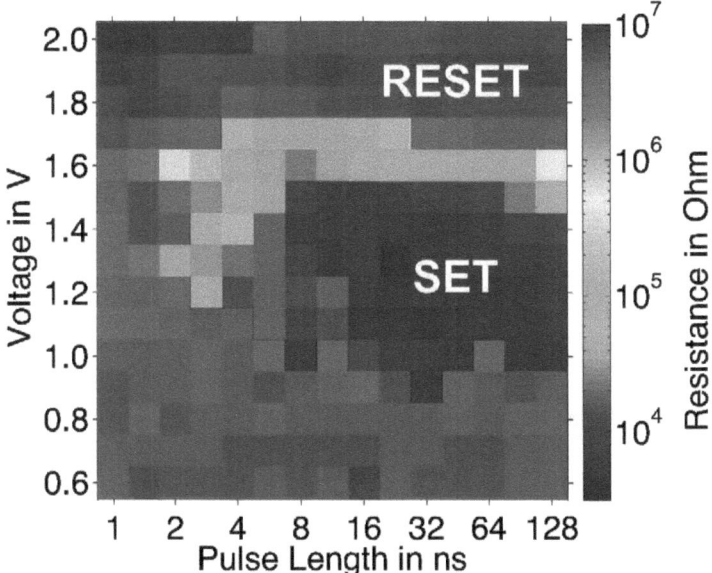

Figure 4.3: PTE diagrams for GeTe. Cell resistance after application of SET pulses with different amplitudes and lengths, each starting from the amorphous RESET state. The color of a data point represents the cell resistance after the test pulse.

increased in a logarithmic way, starting from 1 ns up to 128 ns following the formula: 2^x with $x = 0:0.5:7$ (see footnote[4]), with leading and trailing edges of 2 ns each.

The second part of the experiment investigated four different RESET states, which were obtained by changing the pulse height of the amorphization pulse from 1.8 V to 1.9 V, 2.0 V, and 2.2 V. The increment size of both pulse height and length was increased, and the higher maximum amorphization pulse height of 2.2 V necessitated a higher maximum crystallization pulse height. Therefore, the scan steps of the test pulses for the second experiment was set to: height = 0.6:0.2:2.2 V and length = 2^t ns with t = 0:1:7.

4.2.2 Crystallization Window

The experimental results, revealing the influence of the applied voltage and pulse length on a defined RESET state, are summarized in Fig. 4.3 presented as a PTE diagram. Each colored

---

[4]Notation for incremental iteration, used in MatLab to display for-loops: x = a:b:c describes an increase from x starting with the value a up to c with an increment of b.

square in the diagram represents the cell resistance after a test pulse of the applied parameters represented by the x- and y-coordinates. The color code's scale is logarithmic to visualize the large resistance changes induced by the test pulses. Due to the high resistivity of GeTe's amorphous phase, the resistance of the RESET state is very high (~ 6 MΩ), indicated by red squares. As a consequence of the high contrast to the resistivity of GeTe's crystalline phase of six orders of magnitude (table 4.2), even very thin layers of amorphous material above the heating electrode induce a significant change of the cell resistance.

The initial state before each test pulse was a RESET state with 6 MΩ cell resistance which was conserved, if the test pulse's height was below the threshold voltage of this state; in this case ~ 0.9 V. If the applied voltage U_a exceeds the threshold voltage U_{th}, and the pulse length is long enough, then the cell resistance will decrease to ~ 3 kΩ, represented by blue squares. In the case that U_a is larger than U_{th}, but the pulse length is shorter than the time needed for a complete crystallization at the induced temperature, the resistance will not reach the minimum, but an intermediate value. A further scenario would be the case that U_a is so large (>1.5 V) that not only U_{th} is exceeded, but also the evoked current is high enough to melt parts of the phase-change material, and causes a re-amorphization.

These constraints lead to a crystallization window of pulse parameters, indicated by blue squares in Fig. 4.3. At the optimum pulse height of around 1.3 V, crystallization starts at pulse lengths of 2.8 ns. The SET state with the lowest resistance is reached for pulses longer than 8 ns. These crystallization times are significant smaller than those reported for laser switching experiments in GeTe [Nakayoshi:1992, Coombs:1995]. Furthermore, the observed crystallization speed exceeds even those announced for electronic devices [Lankhorst:2005, Chen:2006].

These differences in crystallization speed can be explained by the nucleation and growth model.

4.2.3 Growth Dominated Recrystallization

One of the major differences concerning the recrystallization in phase-change memory devices and the first crystallization in amorphous as deposited thin films is the presence of crystalline residuals in contact to the amorphous volume. In both electronic and optical phase-change data storages the RESET process creates an amorphous volume in a crystalline surrounding. Subsequent recrystallization of this volume is affected by the interface to the crystalline material.

As discussed in chapter 2.1, the crystallization process can be described as forming and subsequent growing of crystalline nuclei. To describe the recrystallization process in phase-change devices the presence of a device dependent crystalline surrounding has to be taken into account. At the interface to this surrounding the growth process can start without prior nucleation pro-

Phase Transition in Memory Devices

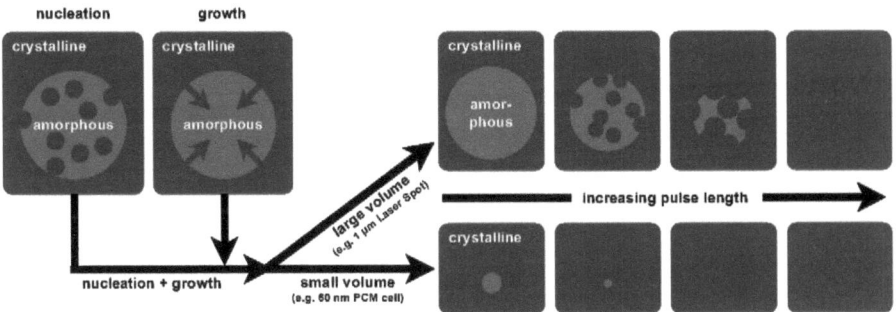

Figure 4.4: Two-dimensional sketch of crystallization in phase-change devices. For larger devices, both nucleation and crystal growth have significant influence on the crystallization time. Smaller volumes can crystallize even before a nucleus has been formed in the switchable volume.

cesses. Materials with low nucleation rates but high growth speeds are assumed to be growth dominated in their recrystallization process [Pieterson:2005]. Their recrystallization times show a strong dependence on the device size. Smaller amorphous volumes can be recrystallized faster by growth from the crystalline rim than large volumes. In contrast, nucleation dominated materials have a high nucleation rate which leads to the formation of crystalline nuclei in the amorphous volume, faster than the crystalline order can grow from the rim the volume. The recrystallization time of a devices build from these materials will not depend on the size of the amorphous volume.

This definition of nucleation and growth dominated materials has been developed for switching experiments with optical laser pulses and amorphous spots with diameters of around 1 µm. However, there is a further aspect which has to be considered: the total size of the amorphous volume. The distinction between nucleation and growth domination is based on relative changes of the size of the amorphous volume, but it is possible to change the dominating mechanism by changing the device size by several orders of magnitude. As sketched in Fig. 4.4 the formation of crystal nuclei within the amorphous volume is important to recrystallize larger volumes. If the switchable volume's size becomes comparable with the nuclei sizes the amorphous material will recrystallize by growth from the surrounding crystalline rim. Therefore, every material should become growth dominated, if the device dimensions become small enough.

Based on this hypothesis, the fast recrystallization times in Fig. 4.3 can be explained by change of the dominating crystallization mechanism. In earlier publications of optical switching experiments [Chen:1986, Huber:1987, Coombs:1995] GeTe was identified as a nucleation dominated phase-change material and crystallization times of around 100 ns were observed. If

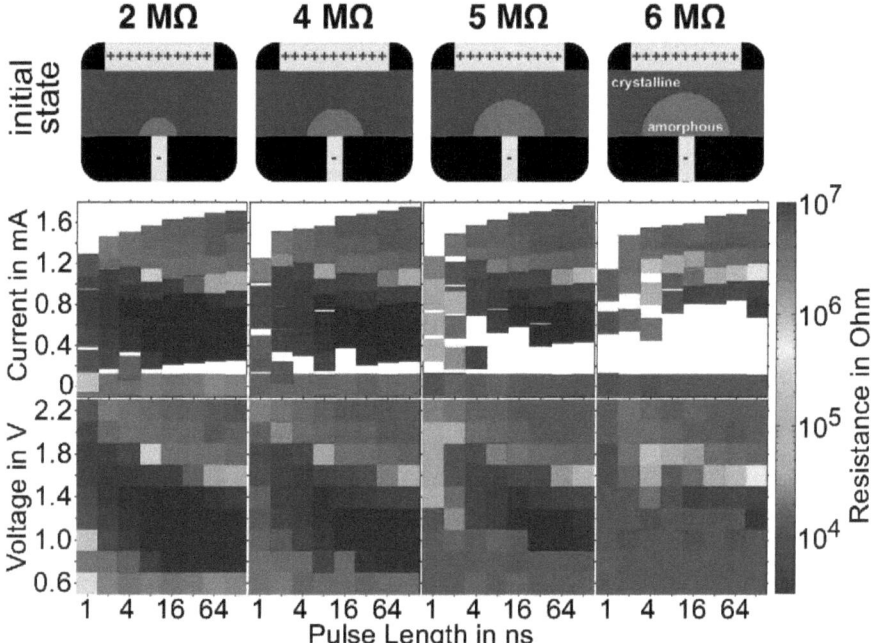

Figure 4.5: PTE diagrams for GeTe. Pulse parameters for crystallization were tested on four different initial states. With increasing size of the initial amorphous volume, the crystallization window shrinks. This proves the domination of the crystal growth mechanism in small phase-change devices.

GeTe memory cells show a growth dominated crystallization mechanism, this would enable recrystallization speeds much faster than those observed in laser experiments which are limited by the nucleation rates. But how to prove this theory?

Besides rescaling the device, in terms of layer thickness or heater diameter, a further possibility to decrease the size of the amorphous volume is the modification of the RESET parameters. A smaller pulse height of the amorphization pulse will induce less heat, and the radius in which phase-change material will melt will become smaller. This will be accompanied by a smaller cell resistance after the RESET. To prove this model, the second part of the experiment (compare section 4.2.1) was performed, which is summarized in Fig. 4.5.

The recrystallization parameters of four different RESET states were investigated. A simplified model of the cell cross-section is sketched in the upper part of Fig. 4.5. The different

Phase Transition in Memory Devices

initial states are indicated by rising resistances resulting from larger amorphous hemispheres (red) above the heating electrode (yellow) in the surrounding crystalline material (blue). Below the cross-sections, the PTE diagrams are plotted which summarize the results of the parameter scan of the corresponding initial state. The upper of two PTE diagrams is the CTR[5], while the lower one is the corresponding VTR[6].

This quiet complex figure contains a lot of information about the behavior of the phase-change material GeTe. The most obvious trend is the shrinking of the crystallization window for higher ohmic initial states in two dimensions. Comparing the threshold voltage U_{th}, which has to be overcome to induce a significant current through the device, there is a clear trend to higher values for higher ohmic initial states. As published by Krebs et al. [Krebs:2009], there is a correlation between U_{th} and the thickness of the amorphous layer. This leads to the interpretation that the applied electric field at which the threshold switch occurs is a characteristic parameter for a phase-change material. Therefore, the increasing threshold voltages in Fig. 4.5 confirm the assumption that the thickness of the amorphous volume correlates with the initial cell resistance.

A correlated, unwanted effect of a higher U_{th} is a loss of information regarding the influence of lower switching currents. As soon as the threshold switch occurs, the phase-change material switches to its amorphous ON state (compare chapters 2.2 and 5), and therefore, the cell resistance plummets to its ON state value ($R_{on} \approx 2\,k\Omega$), which prohibits recording of currents lower than $I = U_{th}/R_{on}$ and leads to white areas in the CTRs.

To investigate the dominating recrystallization process, the time dependence of the crystallization window has to be analyzed. While 1 ns pulse length is sufficient to recrystallize the $2\,M\Omega$-state (Fig. 4.7), the $4\,M\Omega$-state needs at least 2 ns to reach a similar resistance below $10\,k\Omega$, but still shows a significant decrease of resistance within 1 ns. Looking at the results of the $5\,M\Omega$-state, the trend can be confirmed: 4 ns are necessary to SET below $10\,k\Omega$, and the resistance change induced by a 1 ns pulse becomes smaller. Starting from the $6\,M\Omega$-state, there is no resistance change due to a 1 ns pulse at all, and a SET below $10\,k\Omega$ needs at least 8 ns. The time to SET the cell below $10\,k\Omega$, as a criterium for a nearly full SET, will be defined as t_{set}. The trend of t_{set} confirms (Fig. 4.6) a growth dominated recrystallization process in small amorphous volumes surrounded by a crystalline phase.

4.2.4 Competing with DRAM Speed

The observed change of GeTe to become growth dominated at small dimensions could be the breakthrough in the design of ultra-fast phase-change memories: the speed of phase-change

---

[5] Current-Time-Resistance diagram; induced current, pulse plateau length, resistance after pulse.
[6] Voltage-Time-Resistance diagram; applied voltage, pulse plateau length, resistance after pulse.

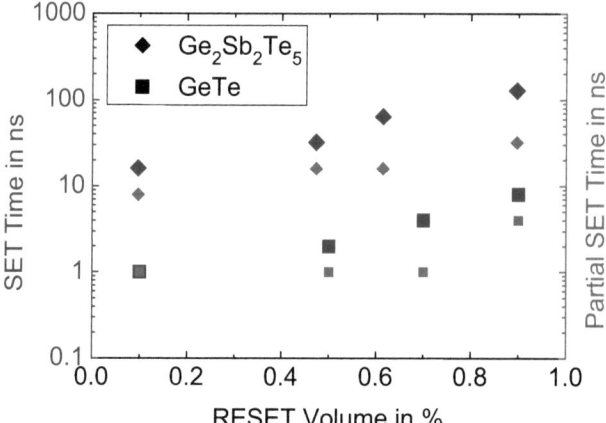

Figure 4.6: SET times for GeTe and $Ge_2Sb_2Te_5$. The SET time has been defined as minimum pulse length necessary to obtain a SET resistance below $10\,k\Omega$. Partial SET times indicate the minimum pulse length necessary to decrease the RESET resistance to a partial SET state. The fraction of amorphous volume was estimated from minimum and maximum RESET resistances. GeTe recrystallizes 16 times faster than $Ge_2Sb_2Te_5$.

memory devices can be increased by shrinking the cell dimensions. Only, the device has to be designed to have a surrounding crystalline rim to enable crystal growth without prior nucleation. Until now, memory manufacturers try to increase the operating speed of phase-change devices by implementing phase-change materials with faster intrinsic crystallization mechanism. Tailoring phase-change alloys and doping of already established candidate materials is considered to be the most promising method to increase switching speeds. However, the results of this chapter could open a new path towards a non-volatile phase-change memory with DRAM speed. Instead of searching a fast nucleating phase-change material, a fast growing material can be used in as small as possible designed memory cells.

Shrinking the cell dimensions is not a new concern of memory manufacturers. To improve storage density, energy consumption, and production costs, there is an ongoing attempt to decrease the minimum feature size (compare technology node[7]) in lithographic procedures to create smaller structures in all silicon based electronic devices. Unfortunately, some technology concepts can not be scaled down as fast as the technology node decreases. The dielectric layers of many field effect transistor (FET) devices must have a certain thickness to prevent device

---

[7] The technology node describes the minimum feature size which can be created on computer chips with the current equipment. The memory chips used in this work were created in a 90 nm line. In 2011, the first companies have introduced 22 nm node.

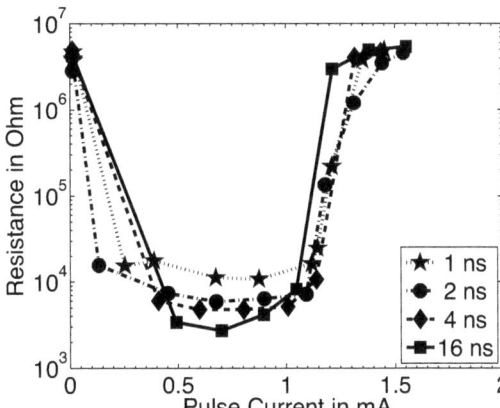

Figure 4.7: Cross-section of Fig. 4.5 through the CTR of the 2 MΩ-state. Cell resistance versus pulse current for applied RESET and SET pulses in the range from 1 to 16 ns.

failure due to unwanted electron tunneling. High-K dielectrics have been used to enable small FETs using the state-of-the-art 22 nm technology node, but it is not certain, if further down scaling is possible. These issues concern flash memory devices, because their storage principle is based on electrons captured in a floating gate isolated by a thin dielectric layer. Other than that, phase-change memory devices do not suffer from future down scaling, but will be improved in at least one of the major critical properties: switching speed.

As presented in section 4.2.3 the crystalline growth mechanism becomes more important in small devices than in larger ones. A simple estimate to describe the critical device size which separates growth dominated recrystallizing devices from nucleation dominated can be the average nucleus size observed in optical switching experiments. When the cell dimensions become less than this average nucleus size[8] every material will become growth dominated, and therefore, the nucleation time becomes negligible. In optical data storage, a fast nucleation mechanism was important for a fast recrystallization of large laser spots. Hence, some phase-change materials with fast growth speeds can be candidate materials for electronic memories, although they have not been used in optical applications, because of their slow nucleation rate, Growth velocities of 30-50 m/s have been reported for several of such candidate materials [Pieterson:2003].

As mentioned at the beginning of this section, GeTe suffers from a strong dependency of its crystallization speed on the precise stoichiometry [Chen:1986, Raoux:2009b]. However, GeTe has the potential to be implemented in electronic memory, if the results of Coombs et al. [Coombs:1995] are considered. They confirm the strong dependency of the nucleation time on stoichiometry, but they reveal an only weak correlation between growth speed and stoi-

---

[8]GeTe shows an average nuclei size of 100 nm in laser experiments [Huber:1987].

Chapter 4

chiometry. Therefore, combined with the results of this chapter, the influence of stoichiometry becomes less important for small, growth dominated electronic memories.

In conclusion this experiment has proven the possibility to recrystallize amorphous phase-change volumes within a few nanoseconds, and therefore, the potential to compete with state-of-the-art DRAM technology in terms of speed. In addition, the observation of crystal growth domination in small devices provides the opportunity to increase the SET speed by scaling the devices.

The following two sections will focus on two of the most established phase-change materials, $Ge_2Sb_2Te_5$ and AgIn-doped Sb_2Te, to support the findings observed for GeTe. Each material represents one class of materials which are dominated by a different recrystallization mechanism. $Ge_2Sb_2Te_5$ is known for its nucleation dominated recrystallization [Coombs:1995], while AgIn-doped Sb_2Te shows a fast growth dominated recrystallization [Lankhorst:2003, Pieterson:2005].

4.3 Switching Experiments with $Ge_2Sb_2Te_5$

Since 1987, $Ge_2Sb_2Te_5$ is used for rewritable optical data storage [Yamada:1987]. Due to its high nucleation rate [Coombs:1995] and its chemical stability, it was, and still is, the most representable material for all phase-change devices. In the field of phase-change device research, there is no second material which has been investigated by so many groups and in such a high number of publications.

The main idea of the experiment described in this section was the attempt to generalize the results obtained from the GeTe experiment. Transferring the conclusions to are more general level could lead to fundamental understanding of the behavior of phase-change materials in electronic devices. Unfortunately, it is not possible just to change the phase-change layer from GeTe to $Ge_2Sb_2Te_5$, and redo the experiment. Some important properties of $Ge_2Sb_2Te_5$ differ from those of GeTe; this necessitates changes of the pulse parameters to test the SET speed. The resistivity of $Ge_2Sb_2Te_5$'s amorphous phase is around one order of magnitude smaller than GeTe's, and both melting and crystallization temperature differ from each other (see table 4.1 and 4.2). Therefore, the pulse parameters of all three initialization pulses had to be changed to create four different initial states with comparable sizes of the amorphous volumes used in the previous experiment.

The results are summarized in Fig. 4.8. By using different scales for voltage, current, and cell resistance, compared to Fig. 4.5, it is possible to obtain a comparable viewgraph. As mentioned before, the exact size of the amorphous volume of the initial state could not be determined, and therefore, comparison of initial states of GeTe and $Ge_2Sb_2Te_5$ with the same

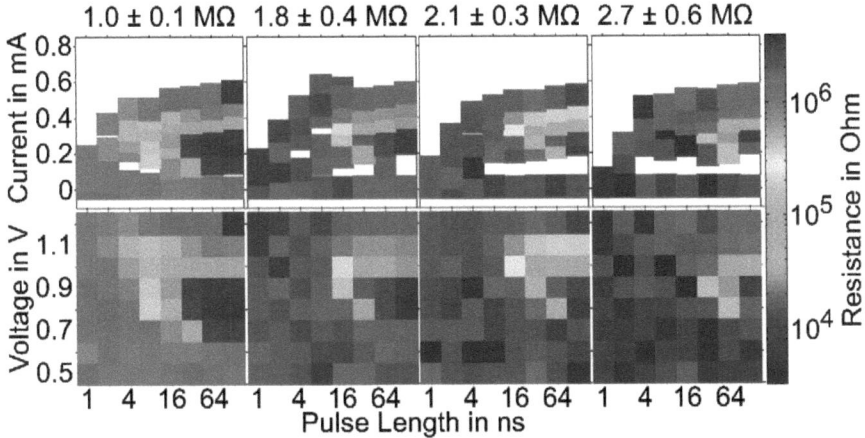

Figure 4.8: PTE diagrams for $Ge_2Sb_2Te_5$. Cell resistance after application of SET pulses with different amplitudes and lengths, each test pulse was applied to the reinitialized amorphous RESET state. The color of a data point represents the cell resistance after the test pulse.

size of the amorphous material can only be done in a qualitative way. To improve the degree of comparability, the maximum possible RESET resistance of both materials was used as a hint what pulse height leads to a full volume RESET of the cell. Additionally, the minimum pulse height for amorphization was determined. This minimum value was defined as the pulse height which was necessary to increases the cell resistance from a SET significantly. Using those two parameter sets, four different initial states from minimum to maximum RESET state were chosen.

At first glance, it is obvious that the crystallization window of $Ge_2Sb_2Te_5$ is much smaller. The timescale of the PTE diagrams has not been changed, and therefore, switching speeds can be compared directly. While the lowest possible RESET state of GeTe could be recrystallized within 1 ns $Ge_2Sb_2Te_5$ can not be recrystallized below 16 ns. Also the highest possible RESET state of GeTe can be recrystallized much faster than the highest initial state of $Ge_2Sb_2Te_5$. Already 4 ns long pulses lead to a change of GeTe cell resistance of two orders of magnitude, while $Ge_2Sb_2Te_5$ does not reach a SET resistance below 64 ns. However, the trend of increasing SET times for initial states of higher resistance can be observed, too. This finding can be understood, if the crystal growth velocity of GeTe is a factor of 16 higher than the one of $Ge_2Sb_2Te_5$.

There is one major difference in the crystallization behavior of $Ge_2Sb_2Te_5$ compared to that

Chapter 4

Table 4.1: Crystallization and melting temperatures of phase-change materials. The crystallization temperatures have been determined by sheet resistance measurements (chapter 3.3), and the melting temperatures have been taken from [a][Kalb:2006], [b][Kloeckner:2007].

	T_{crys} (K)	T_{melt} (K)	T_{crys}/T_{melt} (K)
AgIn-doped Sb_2Te	433	807[a]	0.54
GeTe	423	916[a]	0.46
$Ge_2Sb_2Te_5$	463	985[b]	0.47

of GeTe. Having a closer look at the change of the resistance due to a certain applied voltage, it is noticeable that GeTe shows a sharp transition, while there are intermediate states in the diagram for $Ge_2Sb_2Te_5$. The slower SET mechanism of $Ge_2Sb_2Te_5$ allows partial crystallization of the switchable volume, while the very fast growth of the crystalline order in GeTe leads to an complete SET or no SET on the observed time scale. In addition, the differences in crystallization between GeTe and $Ge_2Sb_2Te_5$ which have been reported by Siegrist [Siegrist:2011] can affect the switching in memory cells to. Siegrist reported intermediate states between amorphous and crystalline order which, among other things, affect the resistivity of phase-change materials. Therefore, it is possible that the amorphous volume in a memory cell crystallizes completely in terms of size, but the atoms achieve only an intermediate state of order.

Further differences between these two materials are compensated through the scaling of the diagrams' axes. On one hand, the lower threshold voltage U_{th} of $Ge_2Sb_2Te_5$ allows switching currents at lower voltages. It is worth mentioning that $Ge_2Sb_2Te_5$ shows a much weaker dependency of U_{th} on the initial state than GeTe. On the other hand, the lower melting temperature T_{melt} of $Ge_2Sb_2Te_5$ allows reamorphization with much smaller currents. Increasing the applied voltage in $Ge_2Sb_2Te_5$ above 1.0 V evokes currents around 0.4 mA, which lead to partial melting and quenching of the switchable volume, and therefore, to cell resistances higher than the minimum SET value. That effect can be observed in GeTe too, but at least 1.6 V are needed to evoke the melting current I_{melt} of around 1.2 mA (experimental results are summarized in table 4.2).

I_{melt} is not proportional to the melting temperature of the material. Material properties like resistivity of the crystalline, or more precisely of the ON state, also have an influence on the induced heat. The Joule heating depends on the applied power $P = R \cdot I^2$, but the reached temperature will be affected by the material's thermal properties like thermal conductivity and heat capacity.

Concerning the different resistivities of the phase-change materials a further axis was rescaled to simplify the comparison with the results of the GeTe cell. Changing the scale of the color

Phase Transition in Memory Devices

Table 4.2: Electrical properties of phase-change materials. Values have been determined using the institute's equipment. Temperature, sample age, and stoichiometric changes due to sample treatment, have a large influence (chapter 2), and therefore, the listed values are not precise measurements, but guidelines that allow comparison of different materials. Values of t_{set} correspond to full RESET initial states.

	ρ_{amo} (Ωcm)	ρ_{crys} (µΩcm)	I_{melt} (mA)	t_{set} (ns)
AgIn-doped Sb$_2$Te	60	1.0	1.0	< 1
GeTe	2000	4.3	1.2	8
Ge$_2$Sb$_2$Te$_5$	1000	85	0.4	128

code helps to visualize the change from RESET to SET. The lower resistivity of Ge$_2$Sb$_2$Te$_5$ was taken into account resulting in a comparable color pattern as shown for GeTe to display the change of state by red to blue colored squares.

In conclusion, the experiment with Ge$_2$Sb$_2$Te$_5$ confirms the observation that increasing SET times for increasing resistances of the initial states can be described using the crystalline growth model. Additionally, it allows to compare important properties of the two phase-change materials.

4.4 Competing Material: AgIn-doped Sb$_2$Te

As the computer market developed, and demanded faster storage devices, the industry began its search for faster crystallizing materials for next generation optical media. Soon AgIn-doped Sb$_2$Te [Iwasaki:1993] was established as a competing material next to Ge$_2$Sb$_2$Te$_5$, but in spite of the technological usage of AgIn-doped Sb$_2$Te, less publication can be found compared with Ge$_2$Sb$_2$Te$_5$. A detailed investigation of AgIn-doped Sb$_2$Te is provided in the phd-thesis of Walter Njoroge [Njoroge:2001]. The fast growth material AgIn-doped Sb$_2$Te [Lankhorst:2003] is a promising candidate for sub-nanosecond switching in phase-change memory devices, and was therefore tested to compare its suitability in technological applications with the one of GeTe and Ge$_2$Sb$_2$Te$_5$.

Hence, an experiment using AgIn-doped Sb$_2$Te phase-change cells was performed as described in chapter 4.2.1. Pulse parameters and the scaling of the PTE diagrams have been modified to simplify the comparison with the results of the previously investigated materials (compare section 4.3). Figure 4.9 summarizes the experimental results.

The low resistivity of the amorphous phase of AgIn-doped Sb$_2$Te complicates the analysis of the switching process. SET and RESET states are identified by measuring the cell resistance

Chapter 4

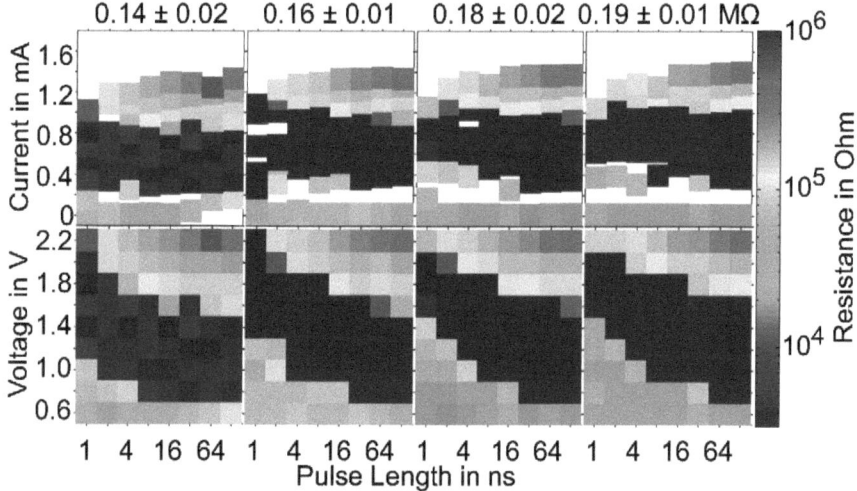

Figure 4.9: PTE diagrams for AgIn-doped Sb$_2$Te. Cell resistance after application of SET pulses with different amplitudes and lengths, each starting from the amorphous RESET state. The color of a data point represents the cell resistance after the test pulse.

subsequently to a switching pulse. A large contrast in resistivity of the crystalline and the amorphous phase is one of the requirements for a large contrast in cell resistances. But in addition, the cell design itself restricts the minimum cell resistance. This was already shown for GeTe in Fig. 4.7. The resistance contrast between a complete SET and a full RESET states is around three orders of magnitude, although the resistivity contrast is around six orders of magnitude.

The minimum cell resistance is limited by the heating electrode to 3 kΩ. A phase-change material will not raise this cell resistance as long as its resistivity is less than $\rho_{\text{limit}} \approx 500\,\mu\Omega\text{cm}$. Therefore, the maximum contrast in cell resistance ΔR_{max} is the ratio of resistivity of the amorphous phase and minimum detectable resistivity:

$$\Delta R_{max} = \frac{\rho_{amo}}{\rho_{limit}} \qquad (4.1)$$

Therefore, compared to GeTe and Ge$_2$Sb$_2$Te$_5$, the lower resistivity of AgIn-doped Sb$_2$Te's amorphous phase decreases the detectable contrast in cell resistance by one order of magnitude.

Nevertheless, a well defined crystallization window can be seen in Fig. 4.9. Remarkable is the fact that the window does not close in terms of pulse lengths. At the same time, the threshold voltage (U_{th}) does rise with increasing initial resistance indicating larger amorphous volumes in higher resistive states. But even the highest resistive initial state can be crystallized with 1 ns short pulses. Extrapolating the crystallization window on the logarithmic timescale leads to the assumption that AgIn-doped Sb_2Te can be switched below 1 ns, and therefore, would be the perfect candidate for a non-volatile memory competing with DRAM in terms of speed.

To summarize this chapter and the findings regarding memory switching in electronic phase-change devices, one aspect is worth emphasizing: recrystallization of amorphous phase-change volumes is possible within 1 ns and even faster. Switching times are limited by the speed of the recrystallization mechanism. In small devices, the crystal growth mechanism dominates the recrystallization. Therefore, down scaling of the device and shrinking the programmable volume will increase the SET speed. A crystalline surrounding has to be provided inside the memory cell to allow fast recrystallization without prior nucleation.

Furthermore, the design of an experiment has been presented which allows comparison of crucial physical properties of phase-change materials. In addition, a method has been developed to analyze complex switching results by visual inspection of PTE diagrams.

Closing the topic of memory switching allows to focus on an effect which was not discussed in detail so far: U_{th} is dependent on the pulse length. This observation is much more pronounced in AgIn-doped Sb_2Te than in the other presented phase-change materials. Known as threshold switching delay time (chapter 2.2), this effect will be the topic of the following chapter 5.

5 Transient Phenomena

In contrast to the phase transition which causes non-volatile changes in phase-change devices, there are also transient phenomena excited by high electric fields. The most common effect is the threshold switching (chapter 2.2) in chalcogenides, first reported by Ovshinsky in the late 1960's [Ovshinsky:1968]. Two aspects of this very important effect will be the topic of the following sections. At first, the transition from the amorphous OFF state to the ON state, the threshold switch itself, will be investigated and discussed in detail (section 5.1). Subsequently, the results regarding the life time of this electronic excitation will be presented in section 5.3. All experiments presented in this chapter have been performed for $Ge_2Sb_2Te_5$ memory cells.

5.1 Threshold Switching Delay Time

Driven by its importance for industrial applications, the threshold switching in phase-change materials has been investigated with increasing effort during the last decade. Especially, the equipment for time resolved measurements in phase-change memory cells has been improved. With increasing signal quality, more details and features of the threshold switching have been identified and have opened new sub topics of investigations. The threshold switching delay time is one of these topics. In 2008, Lavizzari and Karpov [Lavizzari:2008, Karpov:2008] have reported that the threshold switching does not occur instantaneously when the applied voltage on the phase-change cell reaches the threshold voltage. They found a delay time between voltage application and the resistance drop. In addition, they reported a correlation between the height of the applied voltage and the length of the delay time.

Despite the similarities of their findings, Lavizzari and Karpov provide two completely different explanations for the delay in threshold switching (compare chapter 2.2). Furthermore, the observed delay times in both experiments differs from each other. Therefore, an experiment was designed to prove those findings and to investigate the origin of the differences in the observations.

This delay time τ_d can be investigated at best using rectangular pulses, which allow a precise determination of the elapsed time between voltage application and material response, which is detectable as a sudden rise in the current signal. Therefore, an experiment was designed

Chapter 5

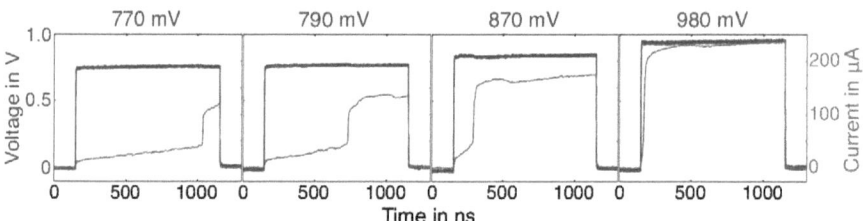

Figure 5.1: Threshold switching delay. Applied voltages: 770 mV, 790 mV, 870 mV, and 980 mV. Dependent on the pulse height, the sudden current rise due to the threshold switch is delayed. Higher voltages of the test pulse decrease the threshold delay time τ_d, until current and voltage edge occur simultaneously.

that should reveal the dependency of τ_d on pulse height and initial state resistance. Using pulses with very steep leading edges of 1 ns, the delay time can be determined very precisely, as presented in section 5.1.1. These results will be complemented by a additional experiment (section 5.1.2), where the leading edge length of the test pulse was varied. In combination, the results of those two experiments characterize the nature of threshold switching in a more profound way, and lay the foundation to compare experimental results of research groups using different experiment designs (section 5.1.3).

5.1.1 Rectangular Pulse Experiment

Four examples of time dependent voltage (applied voltages: 770 mV, 790 mV, 870 mV, and 980 mV) and current measurements of the rectangular test pulses are shown in Fig. 5.1.

Low applied voltages do not lead to an immediate increase of the current simultaneously with the voltage edge. At first, the current increases linearly with time, but after a certain delay time τ_d a steep increase in the current signal is measurable. Carefully re-initializing the cell's RESET state allows to iteratively increase the test pulse's height to determine the dependence of applied voltage and τ_d. The extracted delay times for different initial states are presented in Fig. 5.2. This semilogarithmic plot reveals the dependency of delay times on the height of the applied rectangular pulse. The color of each data point represents the resistance of the initial RESET state.

At first, the discussion of the results will focus on the dark blue data points of a relatively low resistive state (600 kΩ). Due to the fact that the length of the rectangular pulses was 1000 ns, the longest delay time observed in this experiment was limited to this value. The minimum pulse height necessary to induce a threshold switch in this time is 750 mV. Increasing the pulse height leads to a decrease of the observed delay time τ_d. For low voltages, this dependency

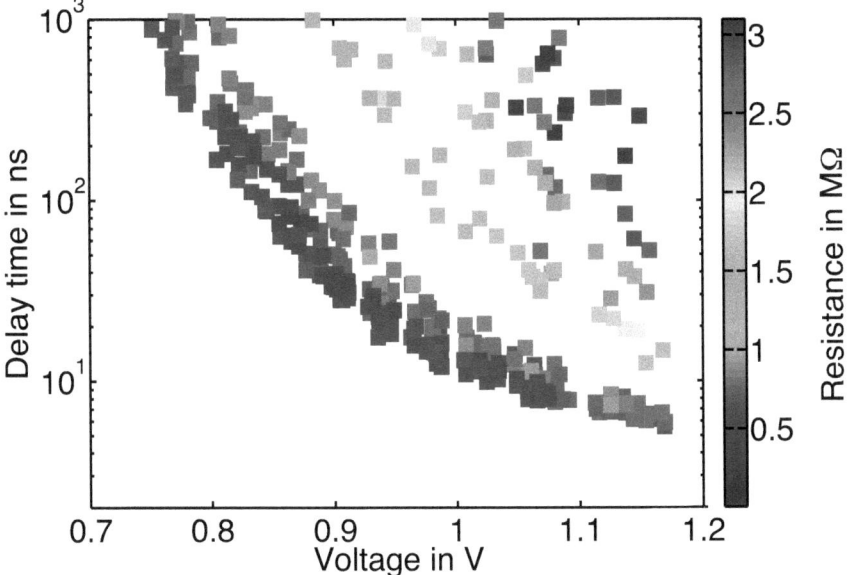

Figure 5.2: Exponential decay of threshold delay time τ_d with increasing applied voltage. The color code represents different initial states on which the rectangular test pulses have been applied. (graphic from [Wimmer:2010])

is very strong. An increase of only 30 % in voltage causes τ_d to decrease by two orders of magnitude.

For higher voltages, the influence of the pulse height on the delay time becomes weaker. The minimum delay time which can be determined by this experiments is 5 ns. Below this limit, the influence of the setup becomes non-negligible and the analysis of the time resolved data becomes inconclusive. Within the limits of the experiment, the correlation of τ_d and applied voltage U_a can be described by an exponential decay as suggested by [Karpov:2008] (compare chapter 2.2.2).

Extending the experiment to investigate the influence of the initial state revealed a dependency on the cell resistance, too. A higher resistance of an initial RESET state has been explained by a larger amorphous volume in the memory cell (see chapter 4). To test the influence of the size of the amorphous volume on the threshold switching, the delay time experiment has been performed for different RESET states indicated by the color code in Fig. 5.2.

RESET states with high initial resistances are difficult to create. In contrast to low RESET resistances, it is necessary to apply very strong pulses to amorphize large volumes in a phase-

Chapter 5

change memory cell. If strong RESET pulses are used, small variations in the SET state can lead to large variations in the subsequent RESET state. Therefore, the distribution of resistances in relation to its mean value is larger for higher resistances (compare Fig. 4.5, 4.8, and 4.9).

This leads to a broader distribution of tested RESET states in this experiment. In contrast to the dark blue data points of the 600 kΩ state and the light blue points of the 900 kΩ state, the measurement sequences of the higher ohmic states do not show a clear curve in the plot. This inhibits a numerical analysis of the dependency of τ_d on the initial state which could be performed by fitting of Karpov's formula 2.20. Hence, this data will be discussed qualitatively at first, before they will be used in a quantitative comparison with the experiment presented in the next section.

For every applied voltage there is a clear influence of the RESET resistance $R_{initial}$ on the delay time. Imaging a vertical line in Fig. 5.2, e.g. through $U_a = 1.03$ V. The data points on this line show the correlation between τ_d and $R_{initial}$. The delay time increases from 10 ns for the lowest RESET state up to 1000 ns for the highest RESET state. A second interesting voltage range is between 1.1 V and 1.2 V. Threshold switching in this range is highly dependent on the length of the test pulses. Small variations of $R_{initial}$ will decide if a test pulse with a typical length of 10 ns to 100 ns will switch a phase-change cell or not.

A further strong dependency of initial state and threshold switching becomes visible by looking at horizontal lines in Fig. 5.2. This dependency of the threshold voltage U_{th} on the thickness of the amorphous layer has already been published in [Krebs:2009]. Krebs reported a linear correlation between U_{th} and layer thickness for several phase-change materials. Therefore, he introduced a threshold field E_{th} as material property. Krebs used a fixed pulse shape to investigate the threshold switching in phase-change memory cells. He uses trapezoidal pulses with fixed edge lengths and varied only the pulse height to test the threshold voltage. For each layer thickness, he increased the pulse height incrementally and measured the cell resistance after each pulse. The height of the applied pulse which leads to a significant drop of the cell resistance was identified as threshold voltage. He could prove that this threshold value depends linearly on the thickness of the amorphous layer. Therefore, he concluded that the applied field strength induces the threshold switching and he was able to determine threshold fields E_{th} for different phase-change materials.

Figure 5.2 shows a comparable result. Assuming a typical test pulse for a threshold voltage scan has a length of 50 ns. This can be visualized as a horizontal line in Fig. 5.2 at $\tau_d = 50$ ns. Data points on this line reveal the correlation between threshold voltage and RESET resistance (indicating layer thickness) for a fixed test pulse length. It is obvious that changing the test pulse length will move the horizontal line, and therefore, will lead to a shift in observed threshold

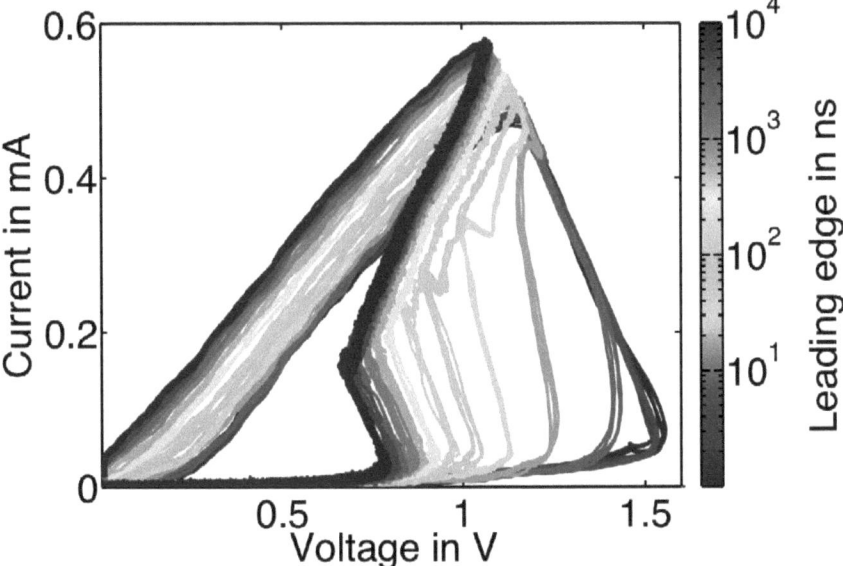

Figure 5.3: Current-voltage diagram of threshold switches during leading edges of different lengths (color code). The high time resolution of 50 ps between two data points leads to quasi continuous lines. The snap back voltage decreases with increasing edge length.

voltages.

Due to the fact that a applied voltage pulse can not be perfectly rectangular but will have a finite leading edge length, it is crucial to investigate the influence of this length on the threshold switching. In the next section a experiment will be presented which investigates this influence.

5.1.2 Leading Edge Variation

Starting from an initial RESET state, $R_{initial} = (1.0 \pm 0.2)$ MΩ, test pulses of 1.6 V with leading edge lengths between 1 ns and 10 µs have been applied to record the threshold switching characteristics. The I-V-curves of all those test pulses are plotted in Fig. 5.3. The color code helps to identify the lengths of the different leading edges used in this experiment. All I-V-curves start from the graph's origin. When the applied voltage increases, the induced current does not increase significantly at first and all colored lines are overlaid and can not been distinguished. At the point where the threshold switch occurs, the current increases suddenly and the voltage will decrease (snap back[1]). After the leading edge segment of the pulse, the maximum voltage of 1.6 V is held for 10 ns, followed by a trailing edge of 10 ns. The I-V-characteristic from this maximum current value is approximately linear and represents the nearly ohmic nature of the amorphous ON state and the SET state.

Due to the threshold delay time, the snap back occurs at lower voltages for longer leading edges. The dark red line of the 10 µs edge shows threshold switching as soon as 0.75 V is reached. Increasing length of the leading edge shifts the threshold voltage up to 1.6 V which is the height of the pulse plateau. This indicates that the threshold switching have occurred after the leading edge segment of the pulse, because the delay time was shorter than the leading edge length.

According to the results of the previous section, the delay time is dependent on the applied voltage. Therefore, a higher pulse plateau value should be used to investigate threshold switching during very short leading edges. But to allow a direct comparison with other leading edge lengths, also those longer pulses have to be set to this higher end value. Unfortunately, a stronger pulse could not be used for longer leading edges, because the induced energy is to high and destroys the cell.

Looking at the starting point of the snap back (see chapter 2.2), it is obvious that a definition of U_{th} using that point does not characterize the material without considering the slope of the voltage rise. On one hand, very steep leading edges delay the snap back to very high voltages. On the other hand, there is a minimum threshold voltage value which can be approximated by the snap back point of increasing longer leading edge. Hence, a minimum field strength, which is necessary to induce a threshold switch at all, can be estimated. Figure 5.4 provides the threshold voltages extracted from Fig. 5.3, using the highest voltage measured at the cell before the snap back happens.

5.1.3 Field Dependent Threshold Time

The result of the two previous sections can be rephrased to be interpreted from a new point of view. On one hand, in section 5.1.1 it has been shown that there is a delay time τ_d which describes the time between application of a voltage and the moment of the threshold switching. The correlation between voltage height and delay time has been shown in Fig. 5.2 using rect-

---

[1] The snap back describes the voltage drop during the threshold switching. The applied voltage on a memory cell is not equal to the voltage drop over the phase-change layer. Due to the resistance of the electrodes, the applied voltage is divided according to the ratio of the resistances of the phase-change layer and the electrodes. The high resistance of the amorphous OFF state leads to a nearly complete drop of the applied voltage across the phase-change layer. Due to the drastic change in resistivity during the threshold switching, a large fraction of the applied voltage drops afterwards across the heating electrode.

angular pulses with discrete pulse heights. On the other hand, in section 5.1.2 the influence of the delay time has been observed indirectly. Leading edges of different lengths allow to observe the threshold switching during a continuous increase of the applied voltage.

It is necessary to translate the findings of section 5.1.1 in a way that they can be used to explain the results of section 5.1.2 quantitatively. Therefore, a new model was developed to describe the correlation of delay time τ_d and applied voltage U_a: The leading edge of a pulse can be interpreted as a sequence of short rectangular pulses with increasing height. Based on the sampling rate of the scope[2], the length of these pulses was set to 50 ps. The next step was to invert the correlation of τ_d and U_a. So far, τ_d was used to describe the time which is necessary to induce a threshold switching at U_a:

$$\tau_d = \tau_d(U_a). \tag{5.1}$$

The main idea of the new model is that this time changes dynamically if U_a increases before τ_d is reached. The resulting time between voltage application and threshold switching can be calculated by summing the fractions S_{U_a} of elapsed time at U_a and delay time at this voltage:

$$S_{U_a} = \frac{t(U_a)}{\tau_d(U_a)}. \tag{5.2}$$

The threshold switch occurs, if the fraction S_{U_a} 1:

$$S_{U_a} \stackrel{!}{=} 1. \tag{5.3}$$

This allows to consider voltage increases during the delay time by assuming that the partial elapsed delay times at different applied voltages can be combined to describe the necessary time till threshold switch occurs. A simple model to do that is to sum up the fractions S_{U_a} until the sum S reaches 1:

$$\sum_{U_a=0}^{U_{max}} S_{U_a} = S \quad \text{with} \quad S = 1 \quad \text{for} \quad U_{max} = U_{th}. \tag{5.4}$$

The applied voltage U_{max} which leads to S = 1 can be interpreted as threshold voltage U_{th}.

Figure 5.4 shows the decrease of U_{th}, obtained from the experiments with varying leading edge. These values can now be compared with the results of the experiments based on rectangular pulses. Therefore, the calculations of U_{th} obtained from equation 5.4 are plotted as well. The two datasets, plotted in Fig. 5.4, display a very similar correlation between U_{th} and the leading edge length of the test pulse. Their common trend reveals a strong dependency for

---
[2]LeCroy SDA 13000 with 20 GS/s, compare chapter 3.2.

Chapter 5

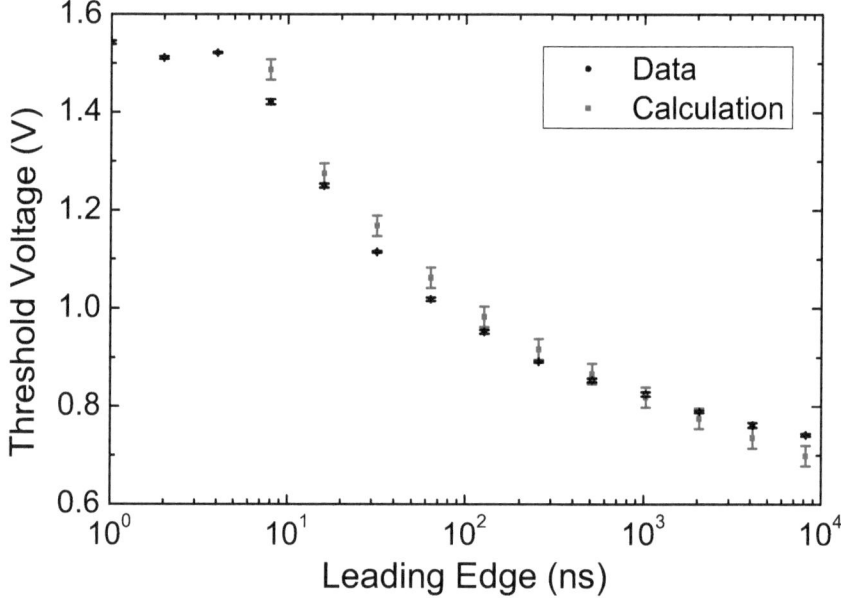

Figure 5.4: Dependency of U_{th} on test pulse shape. Experiments using different leading edges will detect the threshold event at different voltages. Calculations using equation 5.4 allow to transfer measurements of τ_d at different voltages to predict U_{th} for different leading edge lengths. (graphic from [Fleck:2010])

short pulse edges and a decreasing influence of very long leading edges. Extrapolating towards even longer pulse edges motivates the assumption of a minimum threshold voltage for quasi static electrical fields. To prove this assumption, future experiments have to expand the time scale towards longer edge lengths.

In a very strict analysis, both measurements do not agree completely. Regarding the errors of the data points there are deviations between the dataset. These deviations do not affect the common trend and the applicability of the developed model. But they raise the question which differences in the experiments of section 5.1.1 and section 5.1.2 could be responsible for this deviations. One possible explanation could be the slight differences in the initial RESET states. While the initial resistance in section 5.1.1 was $R_{initial}(\text{rectangular}) = (0.9 \pm 0.2)$ MΩ in section 5.1.2 the same initialization procedure has led to $R_{initial}(\text{leading edge}) = (1.0 \pm 0.2)$ MΩ. Another possible explanation could be an insufficient size of the increments used for the calculation. Future experiments should answer these questions.

The results allow to compare switching experiments with different pulse shapes. As mentioned before, there is a strong dependency of U_{th} on the size of the amorphous volume. As reported by Krebs et al. [Krebs:2009], the linear dependency of U_{th} on the thickness leads to a characteristic threshold field E_{th}. To compare his results with future measurements of other groups it is crucial to know the exact method how U_{th} was determined. But even, if the threshold delay effects are considered E_{th} will not be a material property. The time dependency and the field dependency have to be combined to describe the threshold switching of a phase-change material. Therefore, subsequently to the step from voltage to electric field, a second step towards a generalized material property has to be taken. This generalized property could be a field dependent threshold time $\tau_{on}(E_{th})$.

While $\tau_{on}(E_{th})$ describes the excitation of a phase-change material to the amorphous ON state, there is a second characteristic time $\tau_{off}(E_{th})$ necessary to describe the lifetime of this excited state at different applied electric field. The next sections present experiments which tackle this challenge.

5.2 Switching Timescales

There a two timescales which have to be considered during switching in phase-change memory cells. On one hand, the re-crystallization of an amorphous volume in the cell requires the minimum crystallization time τ_{crys}. This re-crystallization time is characteristic for each phase-change material and it depends on the size of the volume as well as on the pulse parameters of the SET pulse. As presented in chapter 4, τ_{crys} can be as short as 1 ns or even less.

On the other hand, the threshold switching delay time τ_d describes the time which is necessary to induce the transition from the amorphous OFF to the amorphous ON state. This transition is mandatory for the re-crystallization in a memory cell. Only the low resistivity of the ON state allows current densities which are high enough to induce sufficient Joule heating for a memory switching. The results of the previous sections have revealed that this delay time τ_d depends on the height of the applied voltages. Delay times between 5 ns and 10 µs have been observed.

Therefore, the orders of magnitude of the timescales τ_{crys} and τ_d are comparable and both depend on the experimental parameters. To investigate effects concerning phenomena of re-crystallization or of threshold switching, the experiment has to be designed carefully to separate the temporal influence of both phenomena.

As long as τ_d is much smaller than τ_{crys}, the investigation of the re-crystallization will not be affected by the threshold delay. If τ_{crys} becomes smaller than τ_d, the precise determination of τ_{crys} becomes difficult. Only time resolved current measurements allow the distinction between

Chapter 5

τ_d and τ_crys. Therefore, further improvement of the measurement technique is mandatory to investigate τ_d and τ_crys below the 1 ns limit.

The fast re-crystallization of phase-change material leads to a further challenge in threshold switching investigations. Ovshinsky discovered the threshold switching in slow crystallizing materials as a reversible effect [Ovshinsky:1968]. He reported a minimum holding voltage for the amorphous ON state. If the applied voltage falls under this value, the material switches back to the amorphous OFF state. A τ_crys of a few nanoseconds, or even less, challenges investigations of the transition from ON to OFF state. The material can crystallizes quasi immediately after the threshold switching, and therefore, the transition to the amorphous OFF state becomes impossible.

In the next section, a experiment will be presented, which was designed to investigate the ON to OFF transition by using a pump and probe method which prevents crystallization of the fast switching material $Ge_2Sb_2Te_5$.

5.3 Excitation Livetime

Ovshinsky reported a transition from the excited amorphous ON state back to the OFF state [Ovshinsky:1968] if the applied voltage falls under a minimum holding voltage. Comparable to the threshold switching delay time τ_d, there should be a characteristic time which describes the delay between the point in time when the applied voltage falls and the point when the resistivity increases. Therefore, a experiment was designed to investigate the livetime of the amorphous ON state when the applied voltage is set to zero. To investigate the transition back to the OFF state, the ON state has to be excited first. A re-crystallization of the material due to this excitation has to be prevented. Hence, a pump and probe concept has been developed to measure the excitation lifetime of the amorphous ON state. A very short excitation pulse and a longer test pulse will be applied after different waiting times to scan the livetime of the amorphous ON state.

5.3.1 Experiment Design

The chosen approach to realize such an experiment is based on the results of the previous chapters 4.3 and 5.1.1. On one hand, $Ge_2Sb_2Te_5$ does not crystallize within a few nanoseconds, even when a detectable current flows (as shown in Fig. 4.8), and on the other hand, $\tau_\text{on}(E_\text{th})$ becomes very small, if U_a is chosen large enough (Fig. 5.2).

A set of pulse parameters extracted from these results have been tested to identify the most suitable excitation pulse. As shown in Fig. 5.5 a rectangular pulse of 5 ns length was

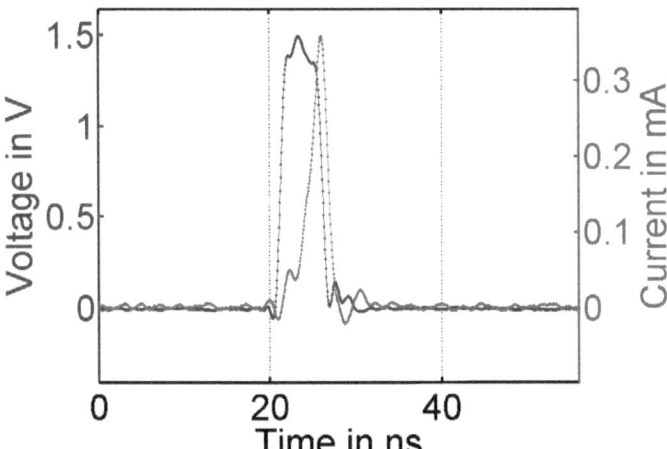

Figure 5.5: Excitation Pulse. Time resolved voltage drop and induced current of a 1.6 V pulse of 5 ns length. The threshold switch occurs at the pulse plateau and forces a current of $(340 \pm 12)\,\mu A$. Pulses with these parameters do not change the initial resistance of more than 5 %.

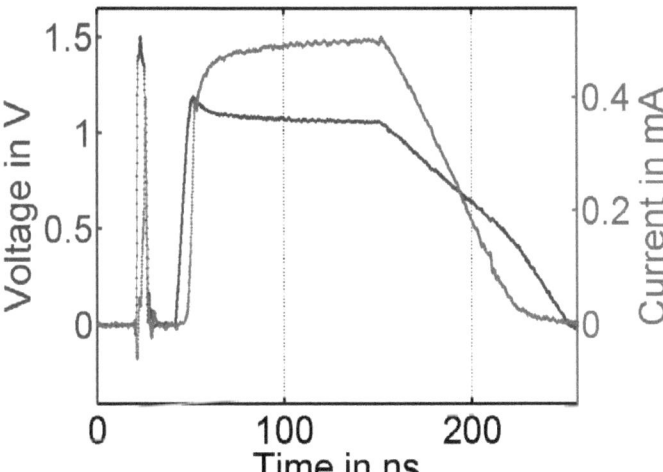

Figure 5.6: Excitation and subsequent probe pulse. The waiting time between pump and probe has been varied to detect its influence on the decrease of U_{th} caused by the excitation pulse. In this example, the OFF state has been recovered partially between excitation and probe pulse. This causes a threshold switching, visible during the leading edge of the probe pulse.

Chapter 5

chosen. The applied voltage was set to 1.6 V, and both leading and trailing edge were set to the minimum value of 1 ns. Preliminary experiments have proven that those pulses do not change the resistance of the initial RESET state by more than 5 %.

A subsequent probe pulse can detect the lifetime of the caused excitation utilizing the change of U_{th}. Assuming that the threshold voltage marks the transition to the amorphous ON state, a subsequent pulse, which will be applied immediately after the first one, will not show a threshold event, because the current will directly follow the I-V-characteristics of the amorphous ON state, and not the one of the OFF state. Adding a waiting time between excitation and probing allows the material to leave the amorphous ON state partially or completely. Hence, also the probe pulse will induce a threshold switching. The threshold voltage of the probe pulse will depend on the elapsed time span between pump and probe pulse. If the OFF state is recovered only partially, the threshold voltage will be smaller than the threshold voltage which can be observed for this probe pulse without prior excitation. If the waiting time is long enough, the probe pulse will show exactly the same behavior as a probe pulse applied directly at the initial state.

The parameters of the pump and probe pulses are based on the results of the previous experiments. From memory switching experiments in chapter 4.3 the minimum re-crystallization times for different pulse heights are know. By combining these constraints with the knowledge of threshold delay times presented in chapter 5.1.1, the parameters of the excitation pulse were chosen. A pump pulse[3] of 5 ns length and leading edges of 1 ns each will excite the amorphous ON state, but will not crystallize the material, if the voltage is set to 1.6 V.

To test the state after the excitation pulse, a probe pulse is applied afterwards. The threshold voltage which can be extracted from the time resolved current measurements of the probe pulse will be used to characterize the cell state. The threshold voltage will be determined using the same techniques which have been used to measure U_{th} to investigate the threshold switching delay using pulses with long leading edges (compare section 5.1.2). A probe pulse[4] with a leading edge of 10 ns and a height of 1.4 V will be used to measure the threshold voltage of the partially excited state after the pump pulse. This probe pulse was applied after varying waiting times from 1 ns up to 10 µs.

In a additional test sequence, the leading edge length of the probe was set to 20 ns. This longer leading edge increases the precision of the determination of U_{th} for long waiting times on one hand, but can affect the recovery process of the excited state for short waiting times on the other hand. In combination, the influence of the probe pulse parameter on the excited state can be minimized. A typical pump and probe pulse pattern is shown in Fig. 5.6.

---

[3]Pump pulse parameters: 1.6 V, 1/5/1 ns.
[4]Probe pulse parameters: 1.4 V, 10/100/100 ns.

Transient Phenomena

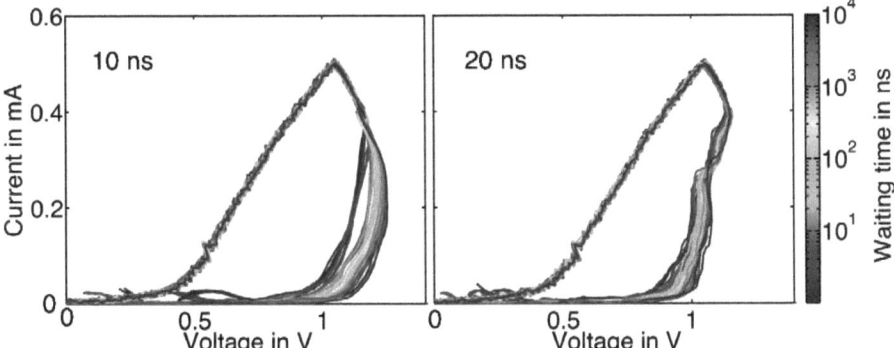

Figure 5.7: Summary of probe pulse I-V-curves. Left: 10 ns leading edge. Right: 20 ns leading edge. The waiting time between pump and probe pulse (indicated by color code) influences the position of the snap back effect caused by the threshold switch. The threshold voltage is decreased by the preliminary excitation pulse.

5.3.2 Threshold Voltage Recovery Time

Figure 5.7 displays the I-V-curves of both kind of probe pulses. The different waiting times utilized between excitation and probing are characterized by a color code, where blue marks the shortest possible waiting time of 1 ns. The figure shows the typical I-V-curves of high voltage pulses applied on amorphous memory cells. No curve shows the expected I-V-curve of a ohmic, metal like amorphous ON state. Even the shortest possible waiting time could not prevent a partial recovery of the amorphous OFF state.

The measured current does not increase significantly with increasing voltage until the threshold voltage is reached. The starting point of the subsequent snap back of the voltage, caused by the sudden increase of the current, is strongly correlated with the length of the elapsed waiting time. After differences in their traces in the snap back segment, both types of probe pulses reach the same maximum current of 0.5 mA, followed by a nearly identical I-V-characteristic of their trailing edge segment.

Significant differences have been recorded during the snap back segment. The test pulses with the 10 ns leading edge show non-sharp starting phase of the snap back. Below 0.35 mA, the I-V-curves after different waiting times can be distinguished easily, before they overlay each other. The starting point of the snap back ranges from 0.9 V to 1.1 V. The test pulses with the longer 20 ns leading edge show a much sharper snap back. Comparable to the I-V-curves of the 10 ns pulses, the traces overlay for currents above 0.4 mA, but can be distinguished below. The starting point of the snap back is shifted to lower voltages and can be located between 0.8 V and 1.0 V.

Both sets of probe pulses prove the idea of a recovery behavior of U_{th} after an excitation. Short waiting times lead to a smaller threshold voltage, while longer waiting times require higher fields to induce threshold switching.

Different numerical methods have been applied determine the threshold voltage from time resolved data. One of the simpler techniques is to define a critical current, e.g. 50 µA, and scan the data for the corresponding voltage value, which is needed to induce this current. This voltage can be identified as a threshold voltage.

Other applied techniques consider the maximum current value reached and focus on certain segments of the snap back, e.g. the segment between 15 % and 50 % of the maximum current reached. A linear fit is applied on this segment. The voltage for which the corresponding current of the fitted line becomes zero is defined to be the threshold voltage. This technique can be varied by using different segments, e.g. 30 % and 50 %.

In this experiment, U_{th} was determined using three different algorithms. These three threshold voltages have been averaged for each waiting time. This value was used to calculate the difference between U_{th} before and after excitation in dependence on the waiting time. The decrease of U_{th} is summarized in Fig. 5.8.

The same logarithmic correlation of waiting time and change of U_{th} is observed for both leading edges. Deviations are observed for very short waiting times, because the influence of the ON state relaxation during the leading edge of the probe pulse can not be neglected on short time scales. An applied electric field during the waiting time will conserve the ON state longer; sufficient high fields (see holding voltage, chapter 2.2) should maintain the ON state forever. Therefore, for waiting times on the order of the leading edge lengths, the measured decrease of U_{th}, is too large. For waiting times much longer than the leading edge lengths, the influence of small holding fields can be neglected.

On the one hand, this experiment does strengthen the model of an electronically excited amorphous state that relaxes with time, if no electric field is applied on the phase-change material anymore. On the other hand, it raises more questions and demands for further investigations. At first, it is obvious that the decrease of U_{th} does not reach the size of (0.77 ± 0.1) V, the U_{th} of the initial RESET state, even for very short waiting times. Furthermore, the investigated timescale was restricted by the design of the measurement software. Unfortunately, it was to too short to detect a complete recovery of U_{th}, and therefore, the time necessary for a full relaxation to the amorphous OFF state is still unknown.

Both open tasks can be handled using the existing equipment, but require fundamental changes of the software and of the experiment design. Instead of determining U_{th} with the help of the leading edge method, a two parameter scan of pulse height and length, according to

Transient Phenomena

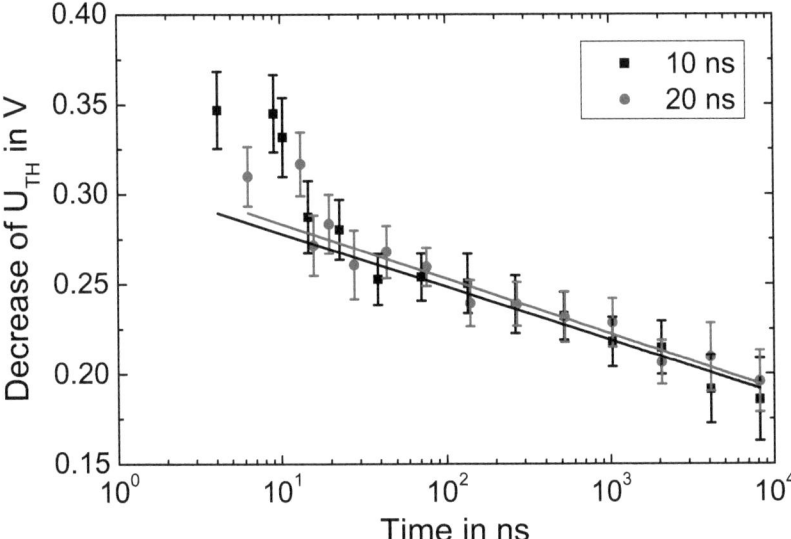

Figure 5.8: Recovery time of threshold voltage after excitation to the amorphous ON state. With increasing waiting times after excitation U_{th} approaches its steady state value following a logarithmic trend. (graphic from [Fleck:2010])

section 5.1.1 using rectangular pulses, can shrink the minimum waiting times towards zero. In addition, re-programming the automatic time scaling of the scope will allow to record events after waiting times of more than 10 µs.

To investigate the influence of the applied field after the excitation, a more elaborate pulse design can be used. Instead of combining rectangular or trapezoidal pulses, the ability of the arbitrary waveform generator could be used to apply small holding voltages after the excitation voltages. A summary of possible experiments, suggested by other authors or extracted from the characteristic predictions of different threshold effect models, can be found in the phd-thesis of Daniel Krebs [Krebs:2010].

In summary, the results of this chapter contribute to a 40 year enduring quest to reveal the nature of the threshold switching effect. Both the excitation and the relaxation of the amorphous ON state have been investigated by taking advantage of the high time resolution of voltage drop and induced current in phase-change cells using the custom made pulsed electrical tester PET. This allowed the investigation of the temporal dependence of the threshold switching effect, and lead to the definition of the field dependent threshold time $\tau_{on}(E_{th})$. In

Chapter 5

addition, first results to determine the relaxation time to the amorphous OFF state have been presented.

Quantitative comparisons with the theoretical models have not lead to results so far. Future analysis of the obtained data and comparison with numerical simulations based on the models should reveal which model describes the threshold switching at best.

These findings complement the investigations of phase-change materials in the amorphous state at timescales of nanoseconds. The next chapter expands the timescale of investigation in the other direction, and measurement techniques will be presented to compare phenomena known from phase-change devices with those in large scale samples in their as deposited amorphous phase.

6 Resistance Drift in Amorphous as Deposited Phase-Change Films

As phase-change materials have been nominated as candidate for future non-volatile storage devices, the focus of investigation shifted from optical towards electronic properties. After the first test devices were created and had been tested successfully, a new effect in melt-quenched phase-change materials was discovered: a continuous increase of the resistance with time, labeled as a resistance "drift" [Pirovano:2004b]. Due to the importance of the resistance for long term data retention in phase-change memory devices, this effect was investigated with increasing effort, and models to describe the physical origin have been elaborated [Ielmini:2007b, Karpov:2007, Boniardi:2009].

Besides the accomplishments to describe the phenomenological observations and to provide numeric models, those publications present explanations from different points of view. They base on experiments performed at phase-change memory cells using melt-quenched amorphous volumes. Experiments in memory cells allow a precise determination of the point in time when the amorphous state is created by a RESET pulse. Therefore, the time dependent change of the resistance can be recorded directly starting from the beginning. A disadvantage is the influence of the memory cell constraints. The amorphous volume is quenched between metal electrodes and it has a large interface to surrounding crystalline material which could affect the drift behavior.

In this work a different approach was chosen. Unstructured thin films of amorphous as deposited phase-change materials were measured in a custom made setup. By this means, the influence of surrounding material can be minimized. In addition, mechanical stress induced by the melt-quenching process can be avoided. This is especially interesting, due to the fact that relaxation of mechanical stress is discussed to be one possible origin of resistance drift. Therefore, a comparison of resistance drift in melt-quenched amorphous volumes in memory cells with drift in unstructured amorphous as deposited thin film can confirm critical aspects of the existing drift models.

Sections 6.2 and 6.3 will present the results of drift experiments with amorphous GeTe. Subsequently, section 6.4 summarizes the results of different phase-change materials and discusses

Chapter 6

the dependence of drift on E_A, the activation energy for carrier conduction.

Details regarding sample geometry and measurement equipment can be found in chapter 3.3. All samples used in this work were measured immediately after the sputter deposition of the phase-change film, or have been stored in nitrogen atmosphere at 5°C until they were mounted in the van-der-Pauw setup.

6.1 Theoretical Background

The resistivities of amorphous phase-change materials measured at room temperature are in the range of $10\,\Omega\text{cm}$ to $2000\,\Omega\text{cm}$ (see table 4.2). Characteristic for semiconductors, their change in resistivity ρ with temperature can be described using the model of thermal activated carrier generation following the Arrhenius equation,

$$\rho(T) = \rho_0 \cdot e^{\frac{E_A}{k_B T}}, \tag{6.1}$$
$$\text{with} \quad \rho_0 = \rho(T = T_0 \neq 0\,\text{K}), \tag{6.2}$$

and k_B the Boltzmann constant. Typical activation energies of phase-change materials are listed in table 6.2. From those values, it becomes obvious that even small deviations of the sample's temperature of $\pm 1°C$ at room temperature will change the material's resistivity by $\pm 5\%$. Therefore, a stable and precise temperature control is indispensable. In addition, it has been observed that the drift mechanism itself is temperature dependent [Pirovano:2004b, Boniardi:2009]. The change of resistivity can be written as

$$R(t) = R_0 \cdot \left(\frac{t}{t_0}\right)^\nu, \tag{6.3}$$
$$\text{with} \quad R_0 = R(t = t_0 \neq 0\,\text{s}), \tag{6.4}$$

with the drift exponent ν which increases with temperature [Boniardi:2009]. Those two temperature dependencies have to be decoupled to distinguish between conductivity change due to carrier generation and drift phenomena. Hence, Boniardi defines two characteristic temperatures: on one hand, the temperature at which the resistance measurement is performed, the read out temperature T_R; on the other hand, the temperature at which the material has drifted, the annealing temperature T_A:

$$R(t, T_A, T_R) = R_0(T_R) \cdot \left(\frac{t}{t_0}\right)^{\nu(T_A, T_R)}, \tag{6.5}$$
$$\text{with} \quad R_0(T_R) = R(t = t_0 \neq 0\,\text{s}, T_R). \tag{6.6}$$

That leads to two kind of experiments, either resistance measurements at a fixed temperature after different drift times at different temperatures ($T_R \neq T_A$), or time resolved resistance measurements during the drifting at different temperatures ($T_R = T_A$). The experiments presented in this chapter belong to the second kind.

The physical origin of the drift is still topic of discussions. Three possible mechanisms have been suggested: structural relaxations, which cause a healing of defects states [Ielmini:2007d], or relaxation of mechanical stress, which either cause a growing distance of trap states [Karpov:2007] or changes the band gap [Pirovano:2004b].

The models which assume a relaxation of mechanical stress within the material, as reason for the change of resistivity with time [Karpov:2007], suggest that stress could have its origin in the process of amorphization. Both sputter deposition and melt quenching in memory cells induce mechanical stress, due to different changes in density within the amorphous material in contrast to the surrounding material during cooling. Authors like Ielmini describe the conduction in disordered semiconductors with a Poole-Frenkel mechanism (compare equation 2.14, chapter 2.2). A relaxation of the mechanical stress would increase the average distance between trap states Δz, and therefore, decrease the conductivity and cause the resistance drift. Ielmini himself suggests [Ielmini:2007d] that structural relaxations heal defects in the disordered material, and therefore, decrease the number of trap states, which will also increase Δz, and cause an increase of resistivity.

In contrast to that idea, Pirovano et al. predict a change of the band gap, resulting from stress relaxation [Pirovano:2004b]. This change of the optical band gap, and its correlation with the activation energy for conduction E_A, has been proven by Krebs [Krebs:2010]. Boniardi developed a physical model to explain the resistance drift as a change of E_A with time, depending on the annealing temperature [Boniardi:2009, Boniardi:2010b]. This model of Boniardi will be the basis for the analysis of the experimental results, and will be presented in the following sections.

6.2 Comparison with Experiments in Memory Devices

At first, the resistance drift of GeTe will be discussed in detail before results of other phase-change materials will be presented in section 6.4. A 100 nm thick layer of amorphous GeTe was deposited by dc-magnetron sputtering, providing a sample geometry as described in chapter 3.3, and was subsequently mounted in the vdP setup at room temperature. After mounting, the sheet resistance of the sample was determined every 30 seconds until demounting. The contact heater design permits to heat the sample to T_A with a slope of 180 K/min. Four different annealing temperatures where used, as shown in Fig. 6.1 and table 6.1.

Chapter 6

Table 6.1: Drift coefficient ν and drift energy β of GeTe for different annealing temperatures. Statistical fluctuations of the temperature control lead to the reported errors. Temperature offsets (<2 K) are not considered in this table.

Temperature (°C)	Drift Coefficient ν	Drift Energy β (meV)
50.000 ± 0.010	0.122 ± 0.002	3.40 ± 0.04
70.000 ± 0.014	0.123 ± 0.002	3.64 ± 0.04
85.000 ± 0.017	0.122 ± 0.002	3.77 ± 0.04
110.000 ± 0.022	0.121 ± 0.002	4.00 ± 0.04

In contrast to experiments in memory devices, where the amorphous state is created by a RESET pulse, the exact age of the amorphous volume is not known in as deposited samples. Therefore, equation 6.5 does not provide enough parameters to fit the resistance change upon drifting. A further parameter τ allows to fit the time elapsed between amorphization and start of the measurements:

$$R(t, T_A, T_R) = R_0(T_R) \cdot \left(\frac{(t+\tau)}{t_0}\right)^{\nu(T_A, T_R)}. \qquad (6.7)$$

Figure 6.1 displays the sheet resistance of four samples at different temperatures ($T_A = T_R$) with different sample ages τ. From these measurements, the drift exponent ν can be extracted for each temperature (see table 6.1). Only minor changes of ν are detectable for different temperatures. The average value of 0.122 is in good agreement with drift coefficients reported for $Ge_2Sb_2Te_5$ in memory cells from 0.06 [Pirovano:2004b] up to 0.21 [Boniardi:2009]. The minor change of ν with temperature observed for GeTe in contrast to $Ge_2Sb_2Te_5$ can be explained by the influence of T_A and T_R which are coupled in our experiment with $T_A = T_R$.

Decoupling T_A and T_R allows to define a drift parameter α that is no longer dependent on the read out temperature. Rewriting Boniardi's finding and emphasizing the energy character of this drift parameter by multiplication with k_B leads to the introduction of the drift energy β:

$$\beta(T_A) = \alpha(T_A) \cdot k_B = \nu(T_A, T_R) \cdot T_R \cdot k_B. \qquad (6.8)$$

Using this drift energy β allows to rewrite Boniardi's key formula which describes the resistance drift as a change of the activation energy E_A with time:

$$E_A(T_A, t) = \beta(T_A) \cdot \ln\left(\frac{t}{t_0}\right) + E_{t0}, \qquad (6.9)$$

Resistance Drift in Amorphous as Deposited Phase-Change Films

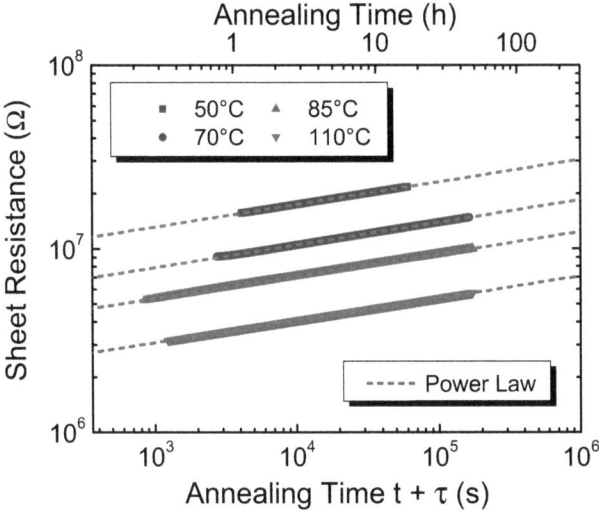

Figure 6.1: Resistance drift in amorphous GeTe films, measured at different temperatures. The drift coefficient shows a only weak dependence on the temperature (for $T_R = T_A$).

$$\text{with} \quad E_{t0} = E_A(t = t_0). \tag{6.10}$$

Therefore, both kinds of resistance change, reversible changes with temperature and irreversible changes with time, have their origin in the same physical parameter E_A: the activation energy E_A increases with time, dependent on the material's temperature. The drift energy β describes the correlation between annealing temperature T_A and drift speed of E_A.

6.3 Dependence of Drift Energy on Temperature

The drift energies β for different temperatures, extracted from Fig. 6.1, are listed in table 6.1. Figure 6.2 reveals the linear dependency of β on the annealing temperature T_A. Hence, GeTe in its amorphous as deposited phase confirms the linear trend observed for $Ge_2Sb_2Te_5$ [Boniardi:2009], and can be expressed as:

$$\beta(T_A) = \beta_S \cdot T_A + \beta_I \tag{6.11}$$
$$\Rightarrow \beta_{GeTe}(T_A) = (0.0099 \pm 0.0005) \frac{\text{meV}}{\text{K}} \cdot T_A + (0.2 \pm 0.2) \,\text{meV}. \tag{6.12}$$

Chapter 6

In contrast to the published data on $Ge_2Sb_2Te_5$ where the trend line can be extrapolated to negative values of β for temperatures below 200 K, formula 6.11 does not predict the possibility of a negative drift behavior. Regarding to formula 6.9, a negative value of β would entail a decrease of E_A with time, and would therefore reverse the resistance drift for large enough times.

It will be a key experiment to check, if the resistance drift is a reversible process, and if materials can be divided into a class of $Ge_2Sb_2Te_5$-like materials where this is possible, and a second class of GeTe-like materials which can not reach negative values of β. Unfortunately, this investigation necessitates resistance measurements at very low temperatures. Below room temperature, the resistivity will increase according to formula 6.1 as long as $k_B T$ is of the order of the energy gap between the Fermi-energy E_F and the conduction band E_C. With decreasing temperature the thermal energy $k_B T$ will become too small to allow thermal activation of carriers and the conduction mechanism will change to a hopping transport between localized states, according to [Mott:1984, Mott:1984b]

$$\rho(T) = \rho_0 \cdot e^{\left(\frac{T_0}{T}\right)^{\frac{1}{4}}} \tag{6.13}$$

The resistivity of the amorphous phase will increase with decreasing temperature causing very high sheet resistances, which will be hard to determine.

Combining equations 6.9 and 6.11 allows to describe the resistance drift using only materials constants, except for one missing parameter: E_{t0}, the activation energy at $t = t_0$. The knowledge of E_{t0} is mandatory to predict E_A after annealing at different temperatures, but due to the definition (formula 6.10) E_{t0} depends on the arbitrary chosen value t_0 and is not a material constant. Despite all improvements in phenomenological and theoretical descriptions of the drift mechanism, the model is still not complete, and it predicts only the changes starting from an arbitrary chosen initial state.

The addition in equation 6.7 which enabled fitting of the experimental data leads to a possible answer to this issue. In summary, equation 6.7 provides one redundant parameter regarding the data fitting: on one hand, t_0 does not contain any physical information, but it is mandatory to divide the elapsed time by another time to get rid of the time's unit, otherwise the powerlaw could not be applied. On the other hand, τ does allow to fit the sample's age but does not solve the problem of units. Redefining t_0 allows to combine both advantages in a single parameter τ:

$$\frac{t+\tau}{t_0} = \frac{t}{\tau} + 1, \tag{6.14}$$
$$\text{with} \quad t_0 = \tau. \tag{6.15}$$

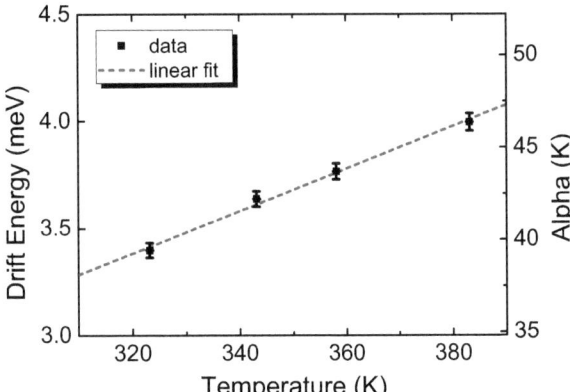

Figure 6.2: Drift energy dependency on annealing temperature.

Therefore, it is necessary to add the sample age τ also to equation 6.9 the same way it has been done in equation 6.7:

$$E_A(T_A, t) = \beta(T_A) \cdot \ln\left(\frac{t}{\tau} + 1\right) + E_0, \quad (6.16)$$

$$\text{with} \quad E_0 = E_A(t = 0). \quad (6.17)$$

Finally, this defines E_0, the materials activation energy at the beginning of the drifting process. In contrast to E_{t0} which was an arbitrary chosen value, the definition of E_0 allows to measure E_0 immediately before the drift experiment at $t = 0$. Therefore, E_0 is a drift experiment independent value which can be determined by other experiments.

In summary the drift phenomenon can be described completely by the knowledge of two intrinsic material constants (β_S, β_I), two experimental parameters (T_A, t), and the start values (τ, E_0):

$$E_A(T_A, t, \tau) = (\beta_S \cdot T_A + \beta_I) \cdot \ln\left(\frac{t}{\tau} + 1\right) + E_0. \quad (6.18)$$

Since T_A and t are set parameters and E_0 can be determined in advance of the experiment, the only unknown parameter which has to be determined by fitting is the age τ of the sample at the beginning of the experiment.

Goal of future experiments should be to determine the two materials constants of different phase-change materials, and search for correlations to structural or electronic properties. The next section 6.4 presents the correlation between drift energy β_{50} measured at the annealing

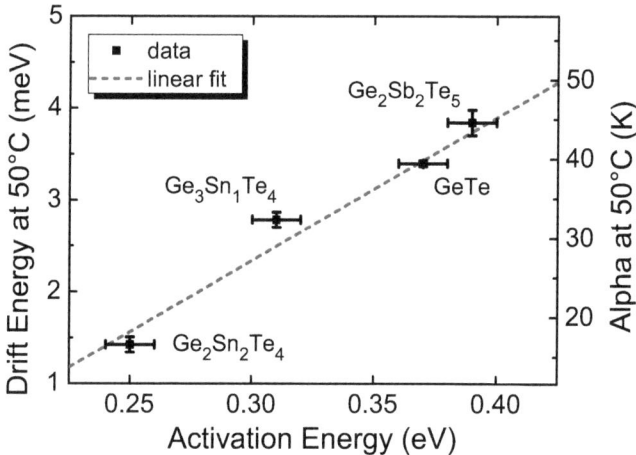

Figure 6.3: Correlation between drift energy measured at 50°C and activation energy for carrier transport at the beginning of the annealing time. (graphic from [Schmidt:2010])

temperature $T_A = 50°C$ and the activation energy E_A at the beginning of the drift experiment.

6.4 Dependence of Drift Energy on Activation Energy for Conduction

Three more materials were investigated to start the systematic search for drift parameters in phase-change materials. $Ge_2Sb_2Te_5$ as the established material in devices was chosen besides two compounds on the pseudobinary line from GeTe to SnTe ($Ge_3Sn_1Te_4$ and $Ge_2Sn_2Te_4$) to investigate possible systematic trends due to the isoelectronic exchange of germanium by tin.

The experimental design was kept as simple as possible. Therefore, only one annealing temperature was used to determine the drift exponent and the drift energy. According to the previous sections, this experiment will not allow to collect the two materials constants (equation 6.18), but it enables a comparison of those four materials at a certain point of the parameter space.

At the beginning of the drift time, the samples have been heated up to $T_A = 50°C$ with 180 K/min. The resistance change during this first 10 seconds of the experiment was used to determine the activation energy E_0 of the material before the drift process according to equation

6.1 for $E_A = E_0$. During the next hours, the sheet resistance was measured and with equation 6.7 the drift coefficient ν_{50} for 50°C was determined. With $T_A = T_R$ and equation 6.8 the drift energy β_{50} was calculated. For all four materials this three parameters are listed in table 6.2.

Figure 6.3 reveals an astonishing correlation: these four materials clearly show a linear dependency of the drift energy β_{50} on the activation energy E_0 immediately before the start of the drift process. Considering the finding of Boniardi that the resistance drift can be explained as a drift of E_A, this new finding strengthens the correlation between drift process and E_A even more. Not only is E_A the value which is changing during the drift process, its start value E_0 seems to be the parameter which actually predicts the speed of the drift process. Therefore, the knowledge of E_0 could predict the complete drift process, if the development of T_A and t are known.

This correlation emphasizes the importance of the modifications and additions of the drift model presented in section 6.3. Future experiments have to prove the predictions resulting from the experiments which were performed in this work and the modified model which describes the drift of the resistivity by changes of the activation energy for electrical conduction.

Summarizing this chapter concerning drift in amorphous phase-change materials, three main results have to be named.

At first, it could be shown that the resistance drift which was discovered in melt-quenched amorphous volumes within phase-change memory can be studied on flat films of amorphous as deposited layers. The custom made setup (chapter 3.3) enables the necessary precise measurements of very high sheet resistances at very stable temperatures which can be achieved with fast heating and cooling rates.

In addition, the observed drift parameters are not only in good agreement with previous published data on phase-change memory cells, they also confirm the existing model of Boniardi, and have allowed modifications to generalize this model.

At last, a new correlation between drift energy and activation energy has been found. This finding suggests to describe the drift as a change of the activation energy of conduction dependent on the start value of E_A at the beginning of the drift process.

Chapter 6

Table 6.2: Correlation between drift coefficient ν_{50}, drift energy β_{50} at 50°C, and activation energy E_A for carrier transport at the beginning of the annealing time for different materials.

	ν_{50}	β_{50} (meV)	E_A (eV)
GeTe	0.122 ± 0.002	3.40 ± 0.04	0.37 ± 0.01
Ge$_2$Sb$_2$Te$_5$	0.138 ± 0.002	3.84 ± 0.14	0.39 ± 0.01
Ge$_3$Sn$_1$Te$_4$	0.100 ± 0.002	2.78 ± 0.09	0.31 ± 0.01
Ge$_2$Sn$_2$Te$_4$	0.051 ± 0.001	1.43 ± 0.09	0.25 ± 0.01

7 Conclusion

Electronic switching in phase-change materials is a complex and interdisciplinary field of research, which origins can be dated back to 1968. Established physical models describe the crystalline phases, and they predict the transitions between crystalline and amorphous phases. But the models for transport phenomena and structural changes within the amorphous phase are still under development and need constructive feedback from experiments. Three aspects concerning characteristics of phase-change material critical for implementation in electronic data storage have been studied in this work: memory switching, threshold switching, and resistance drift.

The speed limits of recrystallization processes in phase-change memory devices have been investigated in state-of-the-art memory cells using a custom made pulsed electrical tester. SET processes within one nanosecond have been observed in two candidate materials, and they could be explained by growth dominated recrystallization. Memory cells with amorphous volumes smaller than the average nucleus size will recrystallize by growth from the surrounding crystalline material. The time for a growth dominated recrystallization of a volume depends on the size of this volume. With future development of the technology node in semiconductor fabrication the minimum structure size, and therefore the cell size, will decrease further. Hence, the switching speed of phase-change memory devices will increase with future technology nodes and it will compete with established volatile memory products in terms of speed.

At one point in time, the switching speed of a phase-change device will not be limited by recrystallization process, but by the threshold switching. Therefore, the crucial times for the transition between amorphous ON and OFF state have been determined, and have led to a change of view in the threshold switching discussion. Early publications describe the threshold effect as a sudden drop in resistivity as soon as a critical voltage U_{th} is reached. Subsequently, U_{th} was replaced by a critical threshold field E_{th}. In addition, a delay time was observed which characterizes the time between the application of a switching voltage and the moment of the resistance drop. In this work, this delay time was investigated and the dependence on applied voltage heights and pulse shape has been shown. This has led to the definition of a field dependent threshold delay time. The threshold switch will occur if the integral of applied electric fields and incubation time will reach a critical value.

These two aspects could be investigated due to the careful and highly specialized design of the pulsed electrical tester PET. Focusing on time resolution and signal quality, the setup is able to test only a few customized low capacitive samples, created in an industrial cooperation, per week, but each cell with high precision.

Those investigations of memory and threshold switching have been complemented by investigations of the resistance drift. While switching between crystalline and amorphous ON and OFF states takes place on the nanosecond time scale, the resistance drift has to be observed on the timescales of hours, weeks and even years. Several publications describe the drift phenomena in phase-change memory cells. The change of resistance and threshold voltages of melt-quenched amorphous volumes have been investigated and have led to phenomenological descriptions of the effect. In addition, theoretical models, based on these data, have been published to explain the origin of the drift. Some models describe the drift as an intrinsic property of amorphous phase-change materials. To test this assumption in this work, the resistance drift was investigated in amorphous as deposited unstructured thin films. Therefore, a second custom made setup has been built to meet the requirements of resistivity measurements at elevated temperatures.

The results obtained from these experiments confirm the data and model published by Boniardi. Boniardi describes the resistance drift as change of activation energy for conduction. His model was developed to describe the drift in memory cells. In this work, Boniardi's model has been modified using results obtained from the experiments with amorphous as deposited films. The model has been improved and generalized to achieve a physical interpretation of all parameters and constants used. The definition of intrinsic material constants allows future experiments to confirm the finding of a new correlation between drift speed and activation energy at the starting point of the drift process.

In summary, three physical phenomena in phase-change memory cells have been studied. The memory switching, describing the phase transition between amorphous and crystalline state, has been investigated using several phase-change materials. A custom made setup and state-of-the-art memory cell have allowed to switch phase-change cells within one nanosecond. This proves suitability of phase-change materials in a non-volatile memory with DRAM-like switching speed. Future experiments have to expand the investigated time scale towards shorter test pulses to identify a minimum pulse length for recrystallization in phase-change memory devices.

Additional investigations of the threshold switching, describing the transition between the amorphous OFF and ON state, have revealed further correlations between switching voltage and delay time. The threshold switch is the consequence of the product of applied voltage

Conclusion

and incubation time. If the integral of this product reaches a critical value, the threshold switch occurs. Future experiments have to quantify these correlations and could led to an exact determination of a field dependent threshold delay time.

Furthermore, the resistance drift in amorphous as deposited unstructured thin films has been investigated. Comparison with experiments performed at phase-change memory cells confirms existing theoretical models. The Boniardi model has been improved to allow a generalized description of resistance drift in both as deposited thin films and melt-quenched memory cells. Future experiments have to prove the correlation between drift speed and activation energy for conduction. In addition, an extrapolation towards low temperatures predicts a reverse drift effect for some materials which has to be proven in cooled resistance measurement setups.

Bibliography

[Adler:1978] D. Adler, H.K. Henisch, and N. Mott. Mechanism of threshold switching in amorphous alloys. *Reviews of Modern Physics*, 50(2):209–220, Apr 1978.

[Adler:1980] D. Adler, M.S. Shur, M. Silver, and S.R. Ovshinsky. Threshold switching in chalcogenide-glass thin-films. *Journal of Applied Physics*, 51(6):3289–3309, Jun 1980.

[Adler:1984] D. Adler, M.S. Shur, M. Silver, and S.R. Ovshinsky. Reply to "Comment on 'Threshold switching in chalcogenide-glass thin-films' ". *Journal of Applied Physics*, 56(2):579–580, Jul 1984.

[Ahn:2010] D.H. Ahn, S.L. Cho, H. Horii, D.H. Im, I.-S. Kim, G.H.Oh, S.O. Park, M.S. Kang, S.W. Nam, and C.H. Chung. PRAM technology: from non volatility to high performances. In *Proceedings of E/PCOS 2010*, volume 9, pages 87–91, Sep 2010.

[Bahl:1970] S.K. Bahl and K.L. Chopra. Amorphous versus crystalline GeTe films. III. Electrical properties and band structure. *Journal of Applied Physics*, 41(5):2196–2212, Apr 1970.

[Baily:2006] S.A. Baily and D. Emin. Transport properties of amorphous antimony telluride. *Physical Review B*, 73(16):165211, Apr 2006.

[Becker:1935] R. Becker and W. Döring. *Ann. Phys.*, 24:719, 1935.

[Bedeschi:2009] F. Bedeschi, R. Fackenthal, C. Resta, E.M. Donze, M. Jagasivamani, E.C. Buda, F. Pellizzer, D.W. Chow, A. Cabrini, G. Calvi, R. Faravelli, A. Fantini, G. Torelli, D. Mills, R. Gastaldi, and G. Casagrande. A bipolar-selected phase change memory featuring multi-level cell storage. *IEEE Journal of Solid-State Circuits*, 44(1):217–227, 2009.

[Bez:2005] R. Bez. Innovative technologies for high density non-volatile semiconductor memories. *Microelectronic Engineering*, 80:249–255, Jan 2005.

[Boniardi:2009] M. Boniardi, A. Redaelli, A. Pirovano, I. Tortorelli, D. Ielmini, and F. Pellizzer. A physics-based model of electrical conduction decrease with time in amorphous Ge2Sb2Te5. *Journal of Applied Physics*, 105(8):084506, April 2009.

[Boniardi:2010b] M. Boniardi, D. Ielmini, S. Lavizzari, A.L. Lacaita, A. Redaelli, and A. Pirovano. Statistics of resistance drift due to structural relaxation in phase-change memory arrays. *IEEE Transactions on Electron Devices*, 57(10):2690–2696, Oct 2010.

[Bruns:2009] G. Bruns, P. Merkelbach, C. Schlockermann, M. Salinga, M. Wuttig, T.D. Happ, J.B. Philipp, and M. Kund. Nanosecond switching in GeTe phase change memory cells. *Applied Physics Letters*, 95(4):043108, Jan 2009.

[Burke:1965] J. Burke. *The Kinetics of Phase Transformations in Metals*. Pergamon Press, Oxford, 1965.

[Burr:2010] G.W. Burr, M.J. Breitwisch, M. Franceschini, D. Garetto, K. Gopalakrishnan, B. Jackson, B. Kurdi, C. Lam, L.A. Lastras, A. Padilla, B. Rajendran, S., and R.S. Shenoy. Phase change memory technology. *Journal of Vacuum Science Technology B*, 28(2):223–262, Jan 2010.

[Chao:2008] D.S. Chao, C. Lien, C.M. Lee, Y.C. Chen, J.T. Yeh, F. Chen, M.J. Chen, P.H. Yen, M.J. Kao, and M.J. Tsai. Impact of incomplete set programing on the performance of phase change memory cell. *Applied Physics Letters*, 92:062108, Feb 2008.

[Chen:1986] M. Chen, K.A. Rubin, and R.W. Barton. Compound materials for reversible, phase-change optical data storage. *Applied Physics Letters*, 49(9):502–504, Jan 1986.

[Chen:2003] Y.C. Chen, C. Chen, C. Chen, J. Yu, S. Wu, S. Lung, R. Liu, and C.Y. Lu. An access-transistor-free (0T/1R) non-volatile resistance random access memory (RRAM) using a novel threshold switching, self-rectifying chalcogenide device. In *International Electron Devices Meeting Technical Digest*, pages 37.4.1–37.4.4, Dec 2003.

[Chen:2006] Y.C. Chen, C.T. Rettner, S. Raoux, G.W. Burr, S.H. Chen, R.M. Shelby, M. Salinga, W.P. Risk, T.D. Happ, G.M. McClelland, M. Breitwisch, A. Schrott, J.B. Philipp, M.H. Lee, R. Cheek, T. Nirschl, M. Lamorey, C.F. Chen, E. Joseph, S. Zaidi, B. Yee, H.L. Lung, R. Bergmann, and C. Lam. Ultra-thin phase-change bridge memory device using GeSb. In *International Electron Devices Meeting Technical Digest*, pages 1–4, 2006.

[Chung:2010] A. Chung, J. Deen, J. Lee, and M. Meyyappan. Nanoscale memory devices. *Nanotechnology*, 21:412001, Sep 2010.

[Coombs:1995] J.H. Coombs, A.P.J.M. Jongenelis, W. van Es-Spiekman, and B.A.J. Jacobs. Laser-induced crystallization phenomena in GeTe-based alloys. II. Composition dependence of nucleation and growth. *Journal of Applied Physics*, 78(8):4918–4928, Jan 1995.

[Detemple:2003] Ralf Detemple. *Strukturelle und kinetische Aspekte der kombinatorischen Materialsynthese am Beispiel der Phasenwechselmedien*. PhD thesis, RWTH Aachen University, Germany, Nov 2003.

[Fleck:2010] Karsten Fleck. *Transiente elektronische Effekte während des Threshold Switches in Phasenwechselmaterialien*. Diploma thesis, RWTH Aachen University, Germany, Dec 2010.

[Friedrich:2000] I. Friedrich, V. Weidenhof, W. Njoroge, P. Franz, and M. Wuttig. Structural transformations of Ge2Sb2Te5 films studied by electrical resistance measurements. *Journal of Applied Physics*, 87(9):4130–4134, Jan 2000.

[Geller:2008] M. Geller, A. Marent, T. Nowozin, D. Bimberg, N. Akcay, and N. Oncan. A write time of 6 ns for quantum dot based memory structures. *Applied Physics Letters*, 92:092108, Mar 2008.

[Gleixner:2007] B. Gleixner, A. Pirovano, J. Sarkar, F. Ottogalli, E. Tortorelli, M. Tosi, and R. Bez. Data retention characterization of phase-change memory arrays. In *45th Annual International Reliability Physics Symposium*, pages 542–546, 2007.

[Happ:2006] T.D. Happ, M. Breitwisch, A. Schrott, J.B. Philipp, M.H. Lee, R. Cheek, T. Nirschl, M. Lamorey, C.H. Ho, S.H. Chen, C.F. Chen, E. Joseph, S. Zaidi, G.W. Burr, B. Yee, Y.C. Chen, S. Raoux, H.L. Lung, R. Bergmann, and C. Lam. Novel one-mask self-heating pillar phase change memory. In *Symposium on VLSI Technology Digest of Technical Papers*, pages 120–121, 2006.

[Henisch:1972] H.K. Henisch, R.W. Pryor, and G.J. Vendura. Characteristics and mechanism of threshold switching. *Journal of Non-Crystalline Solids*, 8-10:415–421, 1972.

[Hermes:2011] C. Hermes, M. Wimmer, S. Menzel, K. Fleck, G. Bruns, M. Salinga, U. Böttger, R. Bruchhaus, T. Schmitz-Kempen, M. Wuttig, and R. Waser. Analysis of transient currents during ultrafast switching of TiO2 nanocrossbar devices. *Electron Device Letters, IEEE*, PP(99):1–3, 2011.

[Huber:1987] E. Huber and E.E. Marinero. Laser-induced crystallization of amorphous gete: A time resolved study. *Physical Review B*, 36(3):1595–1604, Jan 1987.

[Ielmini:2005] D. Ielmini, D. Mantegazza, A.L. Lacaita, A. Pirovano, and F. Pellizzer. Parasitic reset in the programming transient of PCMs. *IEEE Electron Device Letters*, 26(11):799–801, Nov 2005.

[Ielmini:2007] D. Ielmini and Y. Zhang. Evidence for trap-limited transport in the subthreshold conduction regime of chalcogenide glasses. *Applied Physics Letters*, 90(19):192102, Jan 2007.

[Ielmini:2007b] D. Ielmini, A.L. Lacaita, and D. Mantegazza. Recovery and drift dynamics of resistance and threshold voltages in phase-change memories. *IEEE Transactions on Electron Devices*, 54(2):308–315, Feb 2007.

[Ielmini:2007c] D. Ielmini and Y. Zhang. Analytical model for subthreshold conduction and threshold switching in chalcogenide-based memory devices. *Journal of Applied Physics*, 102(5):054517, Sep 2007.

[Ielmini:2007d] D. Ielmini, S. Lavizzari, D. Sharma, and A.L. Lacaita. Physical interpretation, modeling and impact on phase change memory (PCM) reliability of resistance drift due to chalcogenide structural relaxation. In *International Electron Device Meeting*, pages 939–942, Oct 2007.

[Ielmini:2008] D. Ielmini. Threshold switching mechanism by high-field energy gain in the hopping transport of chalcogenide glasses. *Physical Review B*, 78(3):035308, Jul 2008.

[Itri:2004] A. Itri, D. Ielmini, A. Lacaitat, A. Pirovano, E. Pellizzer, and R. Bez. Analysis of phase-transformation dynamics and estimation of amorphous-chalcogenide fraction in phase-change memories. In *42nd Annual International Reliability Physics Symposium*, pages 209–215, Mar 2004.

[Iwasaki:1993] H. Iwasaki, M. Harigaya, O. Nonoyama, Y. Kageyama, M. Takahashi, K. Yamada, H. Deguchi, and Y. Ide. Completely erasable phase change optical disc II: Application of Ag-In-Sb-Te mixed-phase system for rewritable compact disc compatible with CD-velocity and double CD-velocity. *Japanese Journal of Applied Physics 1*, 32(11B):5241–5247, Nov 1993.

[Jung:1996] T.S. Jung, Y.J. Choi, K.D. Suh, B.H. Suh, J.K. Kim, Y.H. Lim, Y.N. Koh, J.W. Park, K.J. Lee, J.H. Park, K.T. Park, J.R. Kim, J.H. Lee, and H.K. Lim. A 3.3 V 128 Mb multi-level NAND flash memory for mass storage applications. In *Solid-State Circuits Conference Digest of Technical Papers*, volume 42, pages 32–33, 412, 1996.

[Kaiser:2004] M. Kaiser, L. van Pieterson, and M.A. Verheijen. In situ transmission electron microscopy analysis of electron beam induced crystallization of amorphous marks in phase-change materials. *Journal of Applied Physics*, 96:3193–3198, Jun 2004.

[Kalb:2005] J.A. Kalb, F. Spaepen, and M. Wuttig. Kinetics of crystal nucleation in undercooled droplets of Sb- and Te-based alloys used for phase change recording. *Journal of Applied Physics*, 98:054910, Sep 2005.

[Kalb:2006] Johannes A. Kalb. *Crystallization kinetics in antimony and tellurium alloys used for phase change recording*. PhD thesis, RWTH Aachen University, Germany, Feb 2006.

[Kao:2010] K.F. Kao, Y.C. Chu, F.T. Chen, M.J. Tsai, and T.S. Chin. Phase-change memory devices operative at 100°C. *IEEE Electron Device Letters*, 31(8):872–874, Aug 2010.

[Karpov:2007] I.V. Karpov, M. Mitra, D. Kau, G. Spadini, Y.A. Kryukov, and V.G. Karpov. Fundamental drift of parameters in chalcogenide phase change memory. *Journal of Applied Physics*, 102(12):124503, Dec 2007.

[Karpov:2008] V.G. Karpov, Y.A. Kryukov, I.V. Karpov, and M. Mitra. Field-induced nucleation in phase change memory. *Physical Review B*, 78(5):052201, Aug 2008.

[Karpov:2008b] I.V. Karpov, M. Mitra, D. Kau, G. Spadini, Y.A. Kryukov, and V.G. Karpov. Evidence of field induced nucleation in phase change memory. *Applied Physics Letters*, 92(17):173501A, Apr 2008.

[Karpov:2008c] V.G. Karpov, Y.A. Kryukov, M. Mitra, and I.V. Karpov. Crystal nucleation in glasses of phase change memory. *Journal of Applied Physics*, 104:054507, Sep 2008.

[Kasap:2004] S. Kasap, J.A. Rowlands, S.D. Baranovskii, and K. Tanioka. Lucky drift impact ionization in amorphous semiconductors. *Journal of Applied Physics*, 96(4):2037–2048, Apr 2004.

[Kau:2009] D.C. Kau, S. Tang, I.V. Karpov, R. Dodge, B. Klehn, J.A. Kalb, J. Strand, A. Diaz, N. Leung, J. Wu, S. Lee, T. Langtry, K.W. Chang, C. Papagianni, J. Lee, J. Hirst, S. Erra, E. Flores, N. Righos, H. Castro, and G. Spadini. A stackable cross point phase change memory. In *International Electron Devices Meeting*, pages 27.1.1–27.1.4, 2009.

[Kim:2005] Y.T. Kim, Y.N. Hwang, K.H. Lee, S.H. Lee, C.W. Jeong, S.J. Ahn, F. Yeung, G.H. Koh, H.S. Jeong, W.Y. Chung, T.K. Kim, Y.K. Park, K.N. Kim, and J.T. Kong. Programming characteristics of phase change random access memory using phase change simulations. *Japanese Journal of Applied Physics 1*, 44(4B):2701–2705, Apr 2005.

[Kim:2009] C. Kim, D. Kang, T.Y. Lee, K.H.P. Kim, Y.S. Kang, J. Lee, S.W. Nam, K.B. Kim, and Y. Khang. Direct evidence of phase separation in Ge2Sb2Te5 in phase change memory devices. *Applied Physics Letters*, 94:193504, May 2009.

[Kloeckner:2007] Daniel Klöckner. Kinetic and Electrical Properties of Chalcogenide Phase Change Alloys. Diploma thesis, RWTH Aachen University, Germany, May 2007.

[Kobayashi:2009] A. Kobayashi and Y. Ogra. Metabolism of tellurium, antimony and germanium simultaneously administered to rats. *Journal of Toxicological Sciences*, 34(3):295–303, Mar 2009.

[Kohary:2011] K. Kohary and C.D. Wright. Electric field induced crystallization in phase-change materials for memory applications. *Applied Physics Letters*, 98:223102, May 2011.

[Krebs:2009] D. Krebs, S. Raoux, C.T. Rettner, G.W. Burr, M. Salinga, and M. Wuttig. Threshold field of phase change memory materials measured using phase change bridge devices. *Applied Physics Letters*, 95(8):082101, Jan 2009.

[Krebs:2009b] D. Krebs, S. Raoux, C.T. Rettner, G.W. Burr, R.M. Shelby, M. Salinga, C.M. Jefferson, and M. Wuttig. Characterization of phase change memory materials using phase change bridge devices. *Journal of Applied Physics*, 106:054308, Sep 2009.

[Krebs:2010] Daniel Krebs. *Electrical transport and switching in phase change materials*. PhD thesis, RWTH Aachen University, Germany, Mar 2010.

[KrusinElbaum:2007] L. Krusin-Elbaum, C. Cabral, K.N. Chen, M. Copel, D.W. Abraham, K.B. Reuter, S.M. Rossnagel, J. Bruley, and V.R. Deline. Evidence for segregation of Te in Ge2Sb2Te5 films: Effect on the "phase-change" stress. *Applied Physics Letters*, 90:141902, Apr 2007.

[Lacaita:2004] A.L. Lacaita, A. Redaelli, D. Ielmini, F. Pellizzer, A. Pirovano, A. Benvenuti, and R. Bez. Electrothermal and phase-change dynamics in chalcogenide-based memories. In *International Electron Devices Meeting Technical Digest*, pages 911–914, 2004.

[Lai:2001] S. Lai and T. Lowrey. OUM - A 180 nm nonvolatile memory cell element technology for stand alone and embedded applications. In *International Electron Devices Meeting Technical Digest*, pages 36.5.1–36.5.4, 2001.

[Lai:2008] S. Lai. Non-volatile memory technologies: The quest for ever lower cost. In *International Electron Devices Meeting*, 2008.

[Lankhorst:2003] M.H.R. Lankhorst, L. van Pieterson, M. van Schijndel, B.A.J. Jacobs, and J.C.N. Rijpers. Prospects of doped Sb-Te phase-change materials for high-speed recording. *Japanese Journal of Applied Physics*, 42(2B):863–868, Feb 2003.

[Lankhorst:2005] M.H.R. Lankhorst, B.W.S.M.M. Ketelaars, and R.A.M. Wolters. Low-cost and nanoscale non-volatile memory concept for future silicon chips. *Nature Materials*, 4:347–352, Mar 2005.

[Lavizzari:2008] S. Lavizzari, D. Ielmini, D. Sharma, and A.L. Lacaita. Transient effects of delay, switching and recovery in phase change memory (PCM) devices. In *International Electron Devices Meeting Technical Digest*, 2008.

[Lavizzari:2010] S. Lavizzari, D. Ielmini, and A.L. Lacaita. Transient simulation of delay and switching effects in phase-change memories. *IEEE Transactions on Electron Devices*, 57(12):3257–3264, Jan 2010.

[Lavizzari:2010b] S. Lavizzari, D. Sharma, and D. Ielmini. Threshold-switching delay controlled by 1/f current fluctuations in phase-change memory devices. *IEEE Transactions on Electron Devices*, 57(5):1047–1054, May 2010.

[Lavizzari:2010c] S. Lavizzari, D. Ielmini, and A.L. Lacaita. A new transient model for recovery and relaxation oscillations in phase-change memories. *IEEE Transactions on Electron Devices*, 57(8):1838–1845, Aug 2010.

[Lee:2009] B.S. Lee, G.W. Burr, R.M. Shelby, S. Raoux, C.T. Rettner, S.N. Bogle, K. Darmawikarta, S.G. Bishop, and J.R. Abelson. Observation of the role of subcritical nuclei in crystallization of a glassy solid. *Science*, 326(5955):980–984, Nov 2009.

[Lencer:2008] D. Lencer, M. Salinga, B. Grabowski, T. Hickel, J. Neugebauer, and M. Wuttig. A map for phase-change materials. *Nature Materials*, 7(12):972–977, Dez 2008.

[Lewis:2009] D.L. Lewis, S. Yalamanchili, and H.H.S. Lee. High performance non-blocking switch design in 3D die-stacking technology. In *IEEE Computer Society Annual Symposium on VLSI*, pages 25–30, 2009.

[Luckas:2010] J. Luckas, D. Krebs, M. Salinga, M. Wuttig, and C. Longeaud. Investigation of defect states in the amorphous phase of phase change alloys GeTe and Ge2Sb2Te5. *Physica Status Solidi C*, 7(3–4):852–856, Mar 2010.

[Matsunaga:2011] T. Matsunaga, N. Yamada, R. Kojima, S. Shamoto, M. Sato, H. Tanida, T. Uruga, S. Kohara, M. Takata, P. Zalden, G. Bruns, I. Sergueev, H.C. Wille, R.P. Hermann, and M. Wuttig. Phase-change materials: Vibrational softening upon crystallization and its impact on thermal properties. *Advanced Functional Materials*, 21(12):2232–2239, Jun 2011.

[Meijer:2008] G.I. Meijer. Who wins the nonvolatile memory race? *Science*, 319:2, Mar 2008.

[Mott:1984] N.F. Mott. *Phil. Mag.*, B19:835, 1984.

[Mott:1984b] N.F. Mott. *The Physics of Hydrogenated Amorphous Silicon Vol. II*, volume 56, page 169. Springer, Berlin Heidelberg, 1984.

[Nakayoshi:1992] Y. Nakayoshi, Y. Kanemitsu, Y. Masumoto, and Y. Maeda. Dynamics of rapid phase transformations in amorphous GeTe induced by nanosecond laser pulses. *Japanese Journal of Applied Physics 1*, 31(2B):471–475, Feb 1992.

[Nardone:2009] M. Nardone, V.G. Karpov, D.C.S. Jackson, and I.V. Karpov. A unified model of nucleation switching. *Applied Physics Letters*, 94:103509, Mar 2009.

[Neale:1970] R. Neale, D. Nelson, and G. Moore. Nonvolatile and reprogrammable, the read-mostly memory is here. *Electronics*, pages 56–60, Sep 1970.

[Nirschl:2007] T. Nirschl, J.B. Phipp, T.D. Happ, G.W. Burr, B. Rajendran, M.-H. Lee, A. Schrott, M. Yang, M. Breitwisch, C.-F. Chen, E. Joseph, M. Lamorey, R. Cheek, S.-H. Chen, S. Zaidi, S. Raoux, Y.C. Chen, Y. Zhu, R. Bergmann, H.-L. Lung, and C. Lam. Write strategies for 2 and 4-bit multi-level phase-change memory. In *International Electron Devices Meeting Technical Digest*, pages 461–464, 2007.

[Njoroge:2001] Walter Njoroge. *Phase Change Optical Recording - Preparation and X-ray Characterization of GeSbTe and AgInSbTe films*. PhD thesis, RWTH Aachen University, Germany, Jan 2001.

[Ohta:1996] T. Ohta. Phase-change optical disk recording systems. In *Summaries of papers presented at the Conference on Lasers and Electro-Optics*, pages 102–103, 1996.

[Okuto:1975] K. Okuto and C.R. Crowell. Threshold energy effects on avalanche breakdown voltage in semiconductor junctions. *Solid State Electron.*, 18(2):161–168, 1975.

[Ovshinsky:1968] S. Ovshinsky. Reversible electrical switching phenomena in disordered structures. *Physical Review Letters*, 21(20):1450–1453, Nov 1968.

[Ovshinsky:1973] S. Ovshinsky and H. Fritzsche. Amorphous semiconductors for switching, memory, and imaging applications. *IEEE Transactions on Electron Devices*, 20(2):91–105, Feb 1973.

[Papandreou:2011] N. Papandreou, H. Pozidis, T. Mittelholzer, G. Close, M. Breitwisch, C. Lam, and E. Eleftheriou. Drift-tolerant multilevel phase-change memory. In *IEEE International Memory Workshop*, volume 3, 2011.

[Park:2007] J.B. Park, G.S. Park, H.S. Baik, J.H. Lee, H. Jeong, and K. Kim. Phase-change behavior of stoichiometric Ge2Sb2Te5 in phase-change random access memory. *Journal of The Electrochemical Society*, 154(3):H139–H141, Jan 2007.

[Pashley:1989] R.D. Pashley and S.K. Lai. Flash memories: the best of two worlds. *IEEE Spectrum*, 26(12):30–33, Dec 1989.

[Pauw:1958] L.J. van der Pauw. A method of measuring specific resistivity and Hall effect of discs of arbitrary shape. *Philips Research Reports*, 13:1–9, Feb 1958.

[Pauw:1958b] L.J. van der Pauw. A method of measuring the resistivity and Hall coefficient on lamellae of arbitrary shape. *Philips Technical Review*, 20:220–224, 1958.

[Pellizzer:2006] F. Pellizzer, A. Benvenuti, B. Gleixner, Y. Kim, B. Johnson, M. Magistretti, T. Marangon, A. Pirovano, R. Bez, and G. Atwood. A 90nm phase change memory technology for stand-alone non-volatile memory applications. In *Symposium on VLSI Technology Digest of Technical Papers*, pages 122–123, 2006.

[Pieterson:2003] L. van Pieterson, M. van Schijndel, J.C.N. Rijpers, and M. Kaiser. Te-free, Sb-based phase-change materials for high-speed rewritable optical recording. *Applied Physics Letters*, 83(7):1373–1375, Jan 2003.

[Pieterson:2005] L. van Pieterson, M.H.R. Lankhorst, M. van Schijndel, A.E.T. Kuiper, and J.H.J. Roosen. Phase-change recording materials with a growth-dominated crystallization mechanism: A materials overview. *Journal of Applied Physics*, 97(8):083520, Jan 2005.

[Pirovano:2003] A. Pirovano, A. Lacaita, A. Benvenuti, F. Pellizzer, S. Hudgens, and R. Bez. Scaling analysis of phase-change memory technology. In *IEEE International Electron Devices Meeting*, pages 29.6.1–29.6.4, Dec 2003.

[Pirovano:2004] A. Pirovano, A. Redaelli, F. Pellizzer, F. Ottogalli, M. Tosi, D. Ielmini, A.L. Lacaita, and R. Bez. Reliability study of phase-change nonvolatile memories. *IEEE Transactions on Device and Materials Reliability*, 4(3):422–427, Sep 2004.

[Pirovano:2004b] A. Pirovano, A.L. Lacaita, F. Pellizzer, S.A. Kostylev, A. Benvenuti, and R. Bez. Low-field amorphous state resistance and threshold voltage drift in chalcogenide materials. *IEEE Transactions on Electron Devices*, 51(5):714–719, May 2004.

[Pirovano:2004c] A. Pirovano, A.L. Lacaita, A. Benvenuti, F. Pellizzer, and R. Bez. Electronic switching in phase-change memories. *IEEE Transactions on Electron Devices*, 51(3):452–459, Mar 2004.

[Porter:1992] D.A. Porter and K.E. Easterling. *Phase Transformations in Metals and Alloys*. Chapman & Hall, London, 1992.

[Prall:2007] K. Prall. Scaling non-volatile memory below 30nm. In *IEEE Non-Volatile Semiconductor Memory Workshop*, volume 22, pages 5–10, Aug 2007.

[Privitera:2004] S. Privitera, E. Rimini, and R. Zonca. Amorphous-to-crystal transition of nitrogen- and oxygen-doped Ge2Sb2Te5 films studied by in situ resistance measurements. *Applied Physics Letters*, 85(15):3044–3046, Aug 2004.

[Raoux:2007] S. Raoux, M. Salinga, J.L. Jordan-Sweet, and A. Kellock. Effect of Al and Cu doping on the crystallization properties of the phase change materials SbTe and GeSb. *Journal of Applied Physics*, 101:044909, Feb 2007.

[Raoux:2008] S. Raoux, J.L. Jordan-Sweet, and A.J. Kellock. Crystallization properties of ultrathin phase change films. *Journal of Applied Physics*, 103:114310, Jun 2008.

[Raoux:2008b] S. Raoux, G.W. Burr, M.J. Breitwisch, C.T. Rettner, Y.C. Chen, R.M. Shelby, M. Salinga, D. Krebs, S.H. Chen, H.L. Lung, and C.H. Lam. Phase-change random access memory: A scalable technology. *IBM Journal of Research and Development*, 52(4-5):465–479, Jul 2008.

[Raoux:2009] S. Raoux, C. Cabral, L. Krusin-Elbaum, J.L. Jordan-Sweet, K. Virwani, M. Hitzbleck, M. Salinga, A. Madan, and T.L. Pinto. Phase transitions in Ge-Sb phase change materials. *Journal of Applied Physics*, 105(6):064918, Jan 2009.

[Raoux:2009b] S. Raoux, H.Y. Cheng, M.A. Caldwell, and H.S.P. Wong. Crystallization times of ge-te phase change materials as a function of composition. *Applied Physics Letters*, 95:071910, Aug 2009.

[Raoux:2009c] S. Raoux, B. Munoz, H.Y. Cheng, and J.L. Jordan-Sweet. Phase transitions in Ge-Te phase change materials studied by time-resolved x-ray diffraction. *Applied Physics Letters*, 95(14):143118, Oct 2009.

[Raoux:2010] S. Raoux, W. Welnic, and D. Ielmini. Phase change materials and their application to nonvolatile memories. *Chemical Reviews*, 110(1):240–267, Feb 2010.

[Redaelli:2004] A. Redaelli, A. Pirovano, F. Pellizzer, A. Lacaita, D. Ielmini, and R. Bez. Electronic switching effect and phase-change transition in chalcogenide materials. *IEEE Electron Device Letters*, 25(10):684–686, Oct 2004.

[Redaelli:2005] A. Redaelli, D. Ielmini, A.L. Lacaita, F. Pellizzer, A. Pirovano, and R. Bez. Impact of crystallization statistics on data retention for phase change memories. *IEDM Technical Digest*, pages 742–745, 2005.

[Redaelli:2005b] A. Redaelli, A.L. Lacaita, A. Benvenuti, and A. Pirovano. Comprehensive numerical model for phase-change memory simulations. In *International Conference on Simulation of Semiconductor Processes and Devices*, pages 279–282, 2005.

[Redaelli:2008] A. Redaelli, A. Pirovano, A. Benvenuti, and A.L. Lacaita. Threshold switching and phase transition numerical models for phase-change memory simulations. *Journal of Applied Physics*, 103:63, Jun 2008.

[Reznik:2007] A. Reznik, S.D. Baranovskii, O. Rubel, G. Juska, S.O. Kasap, Y. Ohkawa, K. Tanioka, and J.A. Rowlands. Avalanche multiplication phenomenon in amorphous semiconductors: Amorphous selenium versus hydrogenated amorphous silicon. *Journal of Applied Physics*, 102:053711, Sep 2007.

[Russo:2009] Ugo Russo. *Physics and device modeling of emerging non volatile memories based on resistance-switching effects*. PhD thesis, Politecnico di Milano, Italy, Nov 2009.

[Salinga:2008] Martin Salinga. *Phase Change Materials for Non-volatile Electronic Memories*. PhD thesis, RWTH Aachen University, Germany, Jun 2008.

[Salinga:2011] M. Salinga and M. Wuttig. Phase-change memories on a diet. *Science*, 332(6029):543–544, Apr 2011.

[Schlockermann:2009] C. Schlockermann, G. Bruns, P. Merkelbach, H. Volker, M. Salinga, and M. Wuttig. Employing advanced characterization tools for the study of phase change materials. In *Proceedings of E/PCOS 2009*, volume 9, page E02, Sep 2009.

[Schlockermann:diss] Carl Schlockermann. *n.n.* PhD thesis, RWTH Aachen University, Germany, to be published.

[Schmidt:2010] Rüdiger Schmidt. Widerstandsdrift in amorphen Phasenwechselmaterialien. Diploma thesis, RWTH Aachen University, Germany, May 2010.

[Sherwin:2001] R.M. Sherwin. Memory on the move. *IEEE Spectrum*, 38(5):55–59, 2001.

[Shibata:2008] N. Shibata, H. Maejima, K. Isobe, K. Iwasa, M. Nakagawa, M. Fujiu, T. Shimizu, M. Honma, S. Hoshi, T. Kawaai, K. Kanebako, S. Yoshikawa, H. Tabata, A. Inoue, T. Takahashi, T. Shano, Y. Komatsu, K. Nagaba, M. Kosakai, N. Motohashi, K. Kanazawa, K. Imamiya, H. Nakai, M. Lasser, M. Murin, A. Meir, A. Eyal, and M. Shlick. A 70 nm 16 Gb 16-level-cell NAND flash memory. *IEEE Journal of Solid-State Circuits*, 43(4):929–937, 2008.

[Shportko:2008] K. Shportko, S. Kremers, M. Woda, D. Lencer, J. Robertson, and M. Wuttig. Resonant bonding in crystalline phase-change materials. *Nature Materials*, 7(8):653–658, Aug 2008.

[Siegrist:2011] T. Siegrist, P. Jost, H. Volker, M. Woda, P. Merkelbach, C. Schlockermann, and M. Wuttig. Disorder-induced localization in crystalline phase-change materials. *Nature Materials*, 10(3):202–208, Jan 2011.

[Sontheimer:2008] Tobias Sontheimer. Crystallization kinetics and characterization of defects in chalcogenide phase change alloys. Diploma thesis, RWTH Aachen University, Germany, Jan 2008.

[Turnbull:1969] D. Turnbull. Under what conditions can a glass be formed? *Contemporary Physics*, 10(5):473–488, Sep 1969.

[Uhlmann:1972] D.R. Uhlmann. A kinetik treatment of glass formation. *Journal of Non-Crystalline Solids*, (7):337, 1972.

[Vollmer:1925] M. Vollmer and A. Weber. *Z. Phys. Chem.*, 119:227, 1925.

[Wang:2005] K. Wang, C. Steimer, D. Wamwangi, S. Ziegler, and M. Wuttig. Effect of indium doping on Ge2Sb2Te5 thin films for phase-change optical storage. *Applied Physics A*, 80:1611–1616, 2005.

[Wang:2008] W.J. Wang, L.P. Shi, R. Zhao, K.G. Lim, H.K. Lee, T.C. Chong, and Y.H. Wu. Fast phase transitions induced by picosecond electrical pulses on phase change memory cells. *Applied Physics Letters*, 93:043121, July 2008.

[Waser:2010] R. Waser, R. Dittmann, M. Salinga, and M. Wuttig. Function by defects at the atomic scale - new concepts for non-volatile memories. *Solid-State Electronics*, 54:830–840, 2010.

[Welnic:2006] W. Welnic, A. Pamungkas, R. Detemple, C. Steimer, S. Blugel, and M. Wuttig. Unravelling the interplay of local structure and physical properties in phase-change materials. *Nat Mater*, 5(1):56–62, Jan 2006.

[Welnic:2007] W. Welnic, S. Botti, L. Reining, and M. Wuttig. Origin of the optical contrast in phase-change materials. *Physical Review Letters*, 98(23):236403, Jun 2007.

[Welnic:2008] W. Welnic and M. Wuttig. Reversible switching in phase-change materials. *Materials Today*, 11(6):20–27, Jun 2008.

[Wengenmayr:2008] R. Wengenmayr. Das Ende des Computer-Alzheimers. *Handelsblatt*, 102:9, May 2008.

[Wimmer:2010] Martin Wimmer. Feldinduzierte Übergangseffekte in Phasenwechselmaterialien. Diploma thesis, RWTH Aachen University, Germany, Nov 2010.

[Woltgens:2003] Han-Willem Wöltgens. *Combinatorial material synthesis applied to Ge-Sb-Te based phase-change materials*. PhD thesis, RWTH Aachen University, Germany, 2003.

[Wong:2010] H.S.P. Wong, S. Raoux, S.B. Kim, J. Liang, J.P. Reifenberg, B. Rajendran, M. Asheghi, and K.E. Goodson. Phase change memory. *Proceedings of the IEEE*, 98(12):2201–2227, Dez 2010.

[Wuttig:2005] M. Wuttig. Phase-change materials - Towards a universal memory? *Nature Materials*, 4(4):265–266, Apr 2005.

[Wuttig:2007] M. Wuttig and N. Yamada. Phase-change materials for rewriteable data storage. *Nature Materials*, 6(11):824–832, Jan 2007.

[Wuttig:2007b] M. Wuttig, D. Luesebrink, D. Wamwangi, W. Welnic, M. Gillessen, and R. Dronskowski. The role of vacancies and local distortions in the design of new phase-change materials. *Nature Materials*, 6(2):122–128, Feb 2007.

[Wuttig:2009] M. Wuttig. Phase change materials: The importance of resonance bonding. *Physica Status Solidi B*, 246(8):1820–1825, Jul 2009.

[Yamada:1987] N. Yamada, E. Ohno, N. Akahira, K. Nishiuchi, K. Nagata, and M. Takao. High speed overwritable phase change optical disk material. *Japanese Journal of Applied Physics*, 26:61–66, Jan 1987.

[Yamada:1991] N. Yamada, E. Ohno, K. Nishiuchi, N. Akahira, and M. Takao. Rapid-phase transitions of GeTe-Sb2Te3, pseudobinary amorphous thin films for an optical disk memory. *Journal of Applied Physics*, 69(5):2849–2856, Mar 1991.

[Yeo:2009] E.G. Yeo, R. Zhao, L.P. Shi, K.G. Lim, T.C. Chong, and I. Adesida. Transient phase change effect during the crystallization process in phase change memory devices. *Applied Physics Letters*, 94:243504, Jun 2009.

[Yeo:2010] E.G. Yeo, L.P. Shi, R. Zhao, K.G. Lim, T.C. Chong, and I. Adesida. Parasitic capacitance effect on programming performance of phase change random access memory devices. *Applied Physics Letters*, 96(4):043506, Jan 2010.

[Yoon:2009] S.M. Yoon, S.Y. Lee, S.W. Jung, Y.S. Park, and B.G. Yu. Enhanced memory behavior in phase-change nonvolatile-memory devices using multilayered structure of compositionally modified Ge-Sb-Te films. *Japanese Journal of Applied Physics*, 48:045502, Apr 2009.

[Zalden:2010] P. Zalden, C. Bichara, J. Van Eijk, C. Braun, W. Bensch, and M. Wuttig. Atomic structure of amorphous and crystallized Ge15Sb85. *Journal of Applied Physics*, 107:104312, May 2010.

[Zhang:2007] T. Zhang, Z. Song, F. Wang, B. Liu, S. Feng, and B. Chen. Advantages of SiSb phase-change material and its applications in phase-change memory. *Appl Phys Lett*, 91:222102, Nov 2007.

i want morebooks!

Buy your books fast and straightforward online - at one of world's fastest growing online book stores! Environmentally sound due to Print-on-Demand technologies.

Buy your books online at
www.get-morebooks.com

Kaufen Sie Ihre Bücher schnell und unkompliziert online – auf einer der am schnellsten wachsenden Buchhandelsplattformen weltweit! Dank Print-On-Demand umwelt- und ressourcenschonend produziert.

Bücher schneller online kaufen
www.morebooks.de

 VDM Verlagsservicegesellschaft mbH
Heinrich-Böcking-Str. 6-8   Telefon: +49 681 3720 174   info@vdm-vsg.de
D - 66121 Saarbrücken   Telefax: +49 681 3720 1749 www.vdm-vsg.de

Printed by Books on Demand GmbH, Norderstedt / Germany